COMPLETE CONTAINER
HERB GARDENING

COMPLETE CONTAINER
HERB GARDENING

Design and Grow Beautiful, Bountiful Herb-Filled Pots

SUE GOETZ

COOL
SPRINGS
PRESS

First Published in 2021 by Cool Springs Press, an imprint of The Quarto Group, 100 Cummings Center, Suite 265-D, Beverly, MA 01915, USA. T (978) 282-9590 F (978) 283-2742 QuartoKnows.com

Cool Springs Press titles are also available at discount for retail, wholesale, promotional, and bulk purchase. For details, contact the Special Sales Manager by email at specialsales@quarto.com or by mail at The Quarto Group, Attn: Special Sales Manager, 100 Cummings Center, Suite 265-D, Beverly, MA 01915, USA.

25 24 23 22 21   1 2 3 4 5

ISBN: 978-0-7603-6779-7

Digital edition published in 2021
eISBN: 978-0-7603-6780-3

Library of Congress Cataloging-in-Publication Data

Goetz, Sue, 1961- author.
Complete container herb gardening : design and grow beautiful,
   bountiful herb-filled pots / Sue Goetz.
Beverly, MA, USA : Cool Springs Press, 2020. | Includes index.
ISBN 9780760367797 (paperback) | ISBN 9780760367803 (ebook)
1. Herb gardening. 2. Container gardening.
LCC SB351.H5 G648 2020 (print) | LCC SB351.H5 (ebook) |
   DDC 635/.7--dc23

LCCN 2020025557 (print) | LCCN 2020025558 (ebook)

Design: Allison Meierding
Cover Image: Christina Salwitz
Page Layout: Allison Meierding
Photography: Sue Goetz, except those by Christina Salwitz (pages
30, 74, 80, 84, 88, 98, 102, 110, 116, 120, 126, 130, 136, 140)

Printed in China

I dedicate this book to Alyssa, Hayley, and Courtney—three beautiful women
in my life who have shared many journeys and support me every step of the way.
Love those Goetz girls!

CONTENTS

9 *Introduction*

13 PART 1
CULTIVATE YOUR HERBAL STYLE

15 *Chapter 1* Choosing the Right Pot for Herbs

31 *Chapter 2* The Perfect Place for Growing Herbs

53 *Chapter 3* Container Garden Design

71 PART 2
GROW THIS! EASY HERB CONTAINER PROJECTS

73 *Chapter 4* Culinary Delights

97 *Chapter 5* Herbal Beverages

109 *Chapter 6* Healing Herbs

125 *Chapter 7* Natural Beauty and Housekeeping

135 *Chapter 8* Herbs for Pollinators

145 PART 3
PLANT AND TEND YOUR HERB CONTAINERS

147 *Chapter 9* Filling and Maintaining the Pots

165 *Chapter 10* Growing New Herbs: Propagation

181 *Resources*

184 *Acknowledgments*

185 *About the Author*

186 *Index*

INTRODUCTION

Medicine, aromatics, flavor, and more. The plants we identify as herbs have so much to give. They offer substances to heal the body, living fragrance, and a harvest of tastes for cooking. The colors and textures of herbs mingled in the landscape may also invite pollinators into the garden. No other group of plants offers as much diversity as herbs. Their usefulness—both in the garden and in our lives—is clear.

My goal in this book is to introduce you to an abundant choice of herbs both historical and modern to grow for their varied usefulness as I show you how to create and care for exciting container herb gardens.

WHAT ARE HERBS?

A simple botanical definition of herbs describes them as soft-stemmed, leafy, herbaceous plants, not necessarily woody like trees. Herbaceous plants die back to the ground every year and can be annual (living one year), biennial (living two years), or perennial (living many years). Herbs are plants that have very distinctive essential oils in their soft plant tissues. These complex oils are what give these plants fragrance, flavor, and perhaps medicinal properties, along with all of the other qualities that lead us to grow them.

To expand your herbal horizons, in this book we're going to explore beyond herbaceous herbs and also consider herbs that are shrubs and trees. Large and distinct woody plants, such as witch hazel (*Hamamelis* spp.) and the true tea plant (*Camellia sinensis*), are highly valued as herbs and lend their tall silhouettes to large potted gardens. Trees and shrubs create layers of interest and provide options for well-rounded and distinct herb gardens in containers.

Discovering both the common and uncommon herbs featured in this book will also give you a new perspective on herbalism and a plant's multiple uses. The plant that flavors your pasta sauce might also be a great host plant for your pollinator garden. Or you may discover a unique flowering evergreen shrub that's also used to make a nurturing tea.

Keep common herbs, such as curly parsley and peppermint, handy in small pots for everyday use.

WHY GROW HERBS IN CONTAINERS?

Smaller homes, apartments, and downsizing all translate to having less space to garden, yet we still can have the delights an herb garden offers. Potted gardens are an easy way to fill our lives with herbs, even in limited space. Besides, herb gardens in containers are among the easiest types of gardens to grow. No weeding for hours, bending over, or even keeping a storehouse of tools to care for it all. Containers are the perfect garden style for busy people and limited spaces.

Growing in any type of container affords the opportunity to plant a garden almost anywhere. It lets you place favorite herbs right where you need them. You do not need acres or even a large garden plot to grow herbs in containers.

Gardens in containers can be tailored and sized to fit on even the tiniest windowsill. If space is a rare commodity and you have room for just a few things, why not choose plants that give back and enrich your life, even if it's just one beloved herb growing on the kitchen counter and harvested for cooking? Or perhaps, if you're lucky, it's many different herbs, crammed into whatever space you have.

Container gardening is all about the ability to fit a garden into your lifestyle. Pottery and other types of containers add a decorative touch, while herbs add their unique fragrance, flavor, and texture to the planting design. For some gardeners, containers are the only option to grow a garden, and for others, pottery becomes a way to lend creativity to larger spaces. Annual summer flowering herbs, such as calendula, potted in color-coordinated containers and tucked into permanent beds give options to change plants with the season.

Gardeners who grow herbs in containers realize their multiple benefits. Many container styles are portable and moveable, which makes a good option for those who rent a home or apartment. Another advantage to growing in containers is the ability to downsize and adapt gardening to changing physical abilities. Pottery size and style choices can be made to accommodate wheelchairs and ease physical movement should the need arise.

The design of this fabulous container gives common rosemary an upscale style for an outdoor living space.

CONTAINER GARDENS FOR ALL

My garden is a mix of ornamental planting borders, raised veggie beds, and a small patch of lawn. It is a pretty typical urban plot. So, why do I love my container gardens when I have all this space to garden? Planted pottery is a good way to have my herbs nearby, making them convenient and accessible. They're right outside the back kitchen door when I am cooking or creating something with herbal ingredients. Glazed, colorful pottery is also a design element I use in my existing planting beds. I tuck bright-colored containers into my gardens and can move them in and out when I want to change things up or add color when other plants are out of season.

In contrast, my children have a different perspective on how they garden. My eldest daughter has a young family, pets, and a home with a small plot of land. A playset, swings, a slide, and a large fire pit for warming up at family gatherings and roasting marshmallows take up the outside space. There is even a small kid-friendly zip line. Her garden consists of a few containers off the back porch because other areas of the landscape are saved for family activity.

My other children live in city apartments with tiny balconies. Pottery, soil, and other garden necessities have to be brought in by stairs or the elevator. Their spaces are just large enough to have seating for morning coffee, but they still share the love of plants passed down from their mom. Containers are tucked all along the edges of their balconies. Their lifestyles define their plant choices; one likes to cook and the other prefers groupings of color and flowers.

You can grow herbs in many different sizes and styles of containers.

YOUR HERBAL JOURNEY

Many times, people ask what herb they should grow. My response is: Grow what you love and will enjoy having in your garden. Most container gardeners do not have a vast amount of room to grow everything, so be picky and grow your favorites. Ask yourself, if you could grow just one or two herbs, what would they be? Once you narrow it down to personal favorites, you can go from there.

There are no rules about how many kinds of herbs make an herb garden complete. Just one plant on your windowsill makes you an herb gardener. One large container filled with a single favorite variety or surrounded by a few other herbs on the same theme—cooking herbs to go with your favorite rosemary; or aromatic plants to go with your favorite lavender—might be all you need.

For anyone considering container gardens, the choices you make are a matter of how growing herbs fits into your life. Start with a plan. Decide what plants you want and get to know their growing habits and what environment makes them thrive. Then look for the appropriate space and decide which type of container will best suit your needs.

In this book, I'll help you develop this plan and more. We'll begin by talking about container selection in chapter 1. Then, in chapter 2, we'll discuss site selection and where to integrate your herb containers into the garden space you have, with information on how to grow herbs indoors. Chapter 3 walks you through designing a stylish and useful container herb garden no matter where you live.

Then I'll share tips for deciding which herbs to nurture. Chapters 4 through 8 explore herb container gardens based on various themes that are easy to assemble and reward-

Fabric wall pockets attached to a solid wood screen create an edible vertical garden with parsley, mints, lavender, kale, and spinach.

ing to grow. Describing culinary combos as well as herbs used for herbal teas, home remedies, household cleaning products, natural beauty products, and pollinator gardens, these chapters show you not just what to grow but also how to harvest and use each herb.

Most herbs do not have tidy, formal habits; they tend to sprawl and spread. The choice of a container, where you place it, and how you put the herbs in a pot will show them at their best.

Chapters 9 and 10 are all about how to procure and plant herbs and how to keep your container herbs looking healthy and attractive.

How does your *container* garden grow?

PART 1
CULTIVATE YOUR HERBAL STYLE

Growing herbs in containers has very few limitations. Success comes from knowing which container is best suited to the growing conditions and size of your area. Then you can create an on-the-spot garden where before none existed. The style and design options offered by potted gardens allow you to find creative ways to make them part of the decor of the surroundings in which you put them.

CHAPTER 1

CHOOSING THE RIGHT POT FOR HERBS

Choosing the right container is as important as selecting the plants to grow in it. What is the best container for growing herbs? There are really no set rules on the type of container to use. You can grow herbs in almost anything that holds soil and allows water to drain away. The best choice comes down to what works for you and for the plant.

Find your container and pick a theme—or maybe the other way around, gather herbs you want to grow and search for containers that suit them. Many container shapes, sizes, and materials are available to gardeners. Let's explore some of them.

CERAMIC POTS

While you don't need to know everything about how a particular ceramic container was made, having some basic knowledge can help you make the best choice.

Clays used in pottery vary from organic ingredients mixed with water to refined blends mixed with additives, such as pigments and binders. All fired clay is considered ceramic. The *type* is based on ingredients added to the clay, the temperature at which the container is processed, and the surface coating. These factors influence how a pot will withstand freezing, thawing, and other variables of weather—good things to consider when you invest in pottery you expect will last.

Plain terra cotta

Long before the nursery industry had plastic pots available, they used terra cotta pots to start plants and grow cuttings to ready them for sale. That earthy aesthetic carried into container garden décor, as herb plants have natural beauty when grown in terra cotta pots.

Besides the look, herbs thrive in this type of pot because of the character of the clay. Terra cotta is a clay that's typically only fired once and at a low temperature. Plain terra cotta rarely has a glaze or coating on it. It is porous and will absorb moisture but also allows excess water to evaporate out of the roots and soil quickly. This material takes on a unique patina from soil salts, moss, or algae, giving terra cotta pots an aged character even if they aren't all that old.

A terra cotta bowl makes a simple tabletop herb garden with chives, parsley, variegated-leaf nasturtiums, and colorful leaf lettuces.

A collection of herbs in smaller terra cotta pots is great for narrow window-sills, shelves, and centerpiece displays on your outdoor dining table.

Terra cotta manufacturing varies, and once you learn to recognize the difference, you will understand how to purchase a quality product. Clay poured into molds is usually less expensive but also can be fragile and break easily. Pressed, molded, and wheel-thrown clays (such as highly desirable Italian terra cotta) tend to be heavier and less fragile because of how they are manufactured and fired.

What type of terra cotta do you need? Less expensive terra cotta pots are good to use in smaller sizes (under 10 inches [25 cm] in diameter). They are easy to use in small herb garden groupings in a basket, in tabletop gardens, or on windowsills. They are also great for starting herb seeds and cuttings. Larger clay pots (12 inches [30.5 cm] in diameter or more) are most likely intended as a permanent feature in your container garden and should be an investment. Well-made terra cotta will age well and last a long time.

Herbs do very well in plain terra cotta pottery due to how easily water flows through terra cotta and evaporates through the porous clay. Containerized herbs need good drainage; root rot is the most common failure in potted gardens. However, the porous nature of clay also means the soil will dry out more quickly than in other types of pottery. Herbs grown in terra cotta need to be watered regularly.

Watering terra cotta

A good watering method for small terra cotta pots that dry out fast is to soak them in a tub of water. Empty pots should be soaked before planting. Soak the entire pot in a tub of fresh, clean water until the clay absorbs the water and turns dark. Watering this way ensures that the dry clay does not wick moisture out of the newly planted soil. If a terra cotta pot is already planted and the soil is very dry and having a hard time soaking up water, place the pot in a tub or stopped sink of fresh water up to the rim of the pot. Hold the pot down until you feel the pot become heavy (very dry pots will try to float).

Algae and mineral deposits on the porous surface of pots add character to terra cotta. The pots will weather naturally, especially when stored undisturbed in a shady, moist environment.

How to clean terra cotta pots

The pore spaces in the clay also make ideal growing places for algae and other undesirable things. Disinfect the inside and remove slime and dirt from the exterior with a quick but gentle scrub with a stiff bristle brush and diluted white vinegar (1 part white household vinegar to 4 parts water). Brushing too hard or using straight vinegar can damage the pot's surface. Rinse well and allow the pot to dry completely before planting.

Winter terra cotta care

The denser the processing of the clay and the heavier and more well made the pot, the better it will withstand freezing weather. The most significant damage to a pot can occur when water-filled pores of the clay freeze and the frozen water expands to a breaking point. Cold climate gardeners living in freezing, wet winters should protect clay pots from the elements.

It's a good idea to bring terra cotta pottery into a frost-free garage or greenhouse to protect it from long periods of below-freezing weather. If containers are fully planted and too heavy to move to a protected area, you can wrap the outside of the pot with insulating plastic bubble wrap. Cover with a second layer of natural fiber burlap tied around to keep the plastic in place (the burlap is a little more attractive than plastic, too). Do not hinder drainage by covering the top or the bottom of the container; the plants still need water and drainage.

A New/Old Look: How to Age Terra Cotta Pots

Terra cotta pots with streaks of white or moss are aged by nature. Minerals such as calcium in the water, soil, and fertilizers come in contact with the clay. The interaction of the clay with weather and elements builds over the years and mottles the surface. Moss and algae also build up and layer on top of the residue, adding more of nature's design. The residue is typically harmless to the plants inside and becomes a desirable look. If you want to age new terra cotta pots, you can speed up the process with either of these methods.

Terra cotta pots can be aged through natural processes or by applying paint or moss.

Aging new pots using paint

You can paint terra cotta pots for an instant color change. Use whites and shades of green.

SUPPLIES

terra cotta pot, clean and completely dry
acrylic paint, white and moss green, if desired
paintbrush or clean paint rag
palette (a plate, pie tin, or other dish works)

Dab a small amount of white paint on a palette. Add water to the paint to create a thin, watery consistency. Use a dry brush technique: Dip a brush or paint rag in the paint and remove as much paint as possible, as you do not want the paint to glob on the pot. Use the damp paintbrush or rag to wipe the pot all around with irregular brush strokes, and leave streaks of the terra cotta color showing through for a natural appearance. The paint will soak quickly into the dry pot. If you want a mossy appearance, dry brush green paint irregularly over some areas of the pot. No need to coat everything. Allow the pot to dry before planting.

Aging new pots using moss

Natural aging can also be done to new pots. Adding organic ingredients that encourage a patina of color to grow in the clay accelerates the aging process. What takes years to occur naturally can be done within a few months in the right conditions.

SUPPLIES

terra cotta pot, clean and dry
plain, unflavored yogurt, or buttermilk
moss or nutrient-rich organic compost

Soak the pot in clean water until the clay darkens. Brush the pot heavily with yogurt or buttermilk until it saturates the clay. Don't worry about covering the whole pot; it will look more natural if you aren't too perfect. Firmly rub the wet, freshly painted areas of the pot with moss; or, if you do not have access to fresh moss, use nutrient-rich organic compost. Turn the pot upside down and store it in a damp area. Spray mist, daily if needed, to keep it moist. In two to four weeks, it should start changing. If you rubbed it with moss, you should see moss bits forming in the pores of the pots. The moss will continue to grow in a moist, shady environment.

You can encourage moss to grow on terra cotta to accelerate aging.

Earthenware and glazed ceramic pottery

Most large glazed garden pottery is basically terra cotta covered in glaze and fired at higher temperatures. The clay is also less porous, resulting in a more substantial pot that will also hold moisture better than plain terra cotta. Clay that is fired at a higher temperature also becomes more durable than a plain clay surface. Another plus when choosing glazed pottery is the diversity of color and styles to choose from.

Here are some things to look for when choosing well-made glazed pottery. Price is usually the first clue. Low-fired pottery poured into molds is often less expensive. Look closely at the texture of the clay inside, where it is not glazed. A well-made, high-fired clay pot should have a smoother texture and heavier, denser weight. Also look for chips, hairline cracks, and irregular glazing before purchasing. All of these flaws will affect the longevity of the pot. Cheaper, less durable pottery tends to break easily. Expensive pottery that is well made is a better investment if you want your container to last and be safe for outdoors year round.

A crucial factor in leaving pots outside in freezing temperatures is that they need to be well draining. You never want the water to collect inside the pot over the winter. Water expands as it freezes and will put pressure on the clay. The pressure can break even the most expensive piece of pottery. Use the same method of wrapping in plastic and burlap as previously described for terra cotta to wrap large ceramic pots too heavy to move to a protected area for winter.

Glazed ceramic pottery comes in many colors and styles.

CONCRETE AND CAST STONE

Concrete and cast stone are basically the same type of material but may differ in their formulations of cement and stone particles. Concrete is more porous than most cast

products, but both types of pottery are heavy and strong. It is unusual for them to break unless dropped. This sturdy material is an excellent choice to withstand harsh elements and freezing weather.

Concrete pots tend to be large and ornate as well as more neutral in color. Most are shades of light gray unless a dye is added to the cement.

Cast stone pots are suitable for windy places or areas where you do not want the container to be moveable. Once filled with soil and plants, they can be very heavy and will stay put. Reminder: Place heavy pottery in its position on a surface that can support its filled weight before adding soil and amendments. It will get heavier with soil and plants and may be too heavy to relocate unless you dismantle it.

A stone bowl filled with a meadow of chamomile.

METAL

Metal containers come in all shapes and sizes. The enduring look and feel of metal can add to a contemporary or traditional design, depending on what type of metal is used. Galvanized, zinc, steel, cast iron—many types of metal are used to fashion containers, including wire for window boxes and hanging baskets. Base your choice on the look you want. Cast iron is heavy and usually ornate. Stainless steel, zinc, and galvanized metals often have thinner walls and can be lightweight, depending on how the pot is made.

Resistant to breakage and freeze damage, metal wears the best of any container material. The downside, however, is that large metal planters gather heat and when placed in the hot sun can warm the soil enough to harm root systems. To help combat this, line the inside of the container with a waterproof, insulating material, such as plastic bubble wrap, or use large grow bags inside. Do not block drainage with your insulating material. You can also plant plastic pots and sink them inside decorative metal pots to keep the soil from coming in contact with the metal.

The most significant disadvantage of metal is that most types, unless treated, will rust and deteriorate over time. Galvanized metal is treated to deter moisture, and other special metals are treated to rust naturally but not break down.

All shapes and sizes of metal containers can be planted with herbs.

These metal urns look heavy but are lightweight and suitable for areas with weight restrictions.

GROW BAGS

Grow bags are a great way to have a container within a container. They help protect the soil from a surface that may become too warm, and they also protect vintage metal and wood containers from soil that may damage them. They are typically made of tightly woven fabric, such as felt, that allows drainage and good air circulation. The bags are sold in many sizes and can be shaped to fit unusually shaped pots.

Grow bags can be used seasonally (stored away for the winter) and are perfect for annual herbs, such as calendula and basil. Small-sized bags are also great for starting seeds. Grow bags contribute to healthy root systems. Plants' roots grow to the edges of the bag and stop; this helps keep plant size in check and also allows healthy, full root systems to take in nutrients and water.

To plant a grow bag, fill one with soil until it is 2 to 3 inches (5 to 8 cm) from the rim, then gently shake the bag until the soil is level and the sides of the bag do not collapse. Plant as you would any other container garden.

Varying sizes of grow bags contain soil and plants help protect the metal in a vintage wheelbarrow.

PLASTIC AND FIBERGLASS

Practicality is usually the reason to pick plastic pots for growing herbs. The pots hold moisture well and don't dry out as fast as natural materials. They tend to be less expensive and unbreakable, too. Quality plastic products can typically be left outside during freezing weather with little risk of damage. The biggest benefit of plastic and fiberglass is that they are much lighter weight than ceramics and metal. The manufacturing process has improved over the years, and many attractive options are available for purchase. Molded plastic is an excellent mimic and can be made to look like real clay, metal, and other forms of pottery.

There are a few downsides to plastic, however. Because plastic is not porous like natural materials, less watering is required, but plastic pots also are prone to overwatering if they do not drain well. Inexpensive molded plastics tend not to hold their shape and can warp or crack in heat or freezing temperatures. Sunlight can also break down plastic over time, making it more brittle and fading its color. High-grade plastics offer a much nicer look and feel.

Fiberglass containers are made from polymer laced with fiberglass strands, making them lightweight and rigid. They're suitable for balconies and rooftops where weight load is a concern, but they're likely to blow over in windy areas.

WOOD

The earthy appearance of wood planters gives the garden a natural look and softens the space. A good natural insulator, wood protects root systems in temperature extremes of both hot and freezing weather. Choose planters made of sustainably sourced redwood, teak, or cedar, which are naturally decay resistant. Softer woods, such as pine, are less expensive but may begin to break down after only a few years. When purchasing or making your own wood planters, avoid the use of treated wood, which may contain harmful chemicals that can leach into soil and your plants.

The worst enemy of wood planters is rot. To help preserve the longevity of a planter, line the inside with a protective material, such as plastic or sheet metal, or use a grow bag to avoid constant soil and moisture contact with the wood.

The rustic look of an old wooden crate or large half barrel as an herb gardening container appeals to many. Garden centers and nurseries carry them, but you might also be able to source used half barrels from distilleries or winemaking facilities. Wherever you find yours, if it's used, be sure to verify its previous contents so nothing dangerous—for instance, cleaning chemicals—leeches into your soil or herbs.

This curly parsley is growing well in a wooden planter.

The structural integrity of most used wooden barrels and crates is such that the container can no longer serve its original purpose; it cannot hold liquid or withstand being packed, shipped, and stacked with other containers. This weakening of the wood shortens the container's life as a planter, so keep that in mind when planting in it. Proper drainage, perhaps by elevating the container slightly with small brick pieces or pot feet, can help extend its life.

My friend David says of this striking planter in his garden: "This fennel has been growing in that same spot and gracing my garden for more than twenty years. Each summer, she grows 9 to 10 feet (2.7 to 3 m) tall and provides an ideal nursery for hundreds of native ladybugs that hatch out and move through the larval, pupae, and adult stages."

Mint is growing beautifully in this recycled piece of enameled cookware.

A vintage-style metal chair takes on a new purpose as an herb planter.

VINTAGE AND REPURPOSED

For a fresh twist on the *recycle, reuse, repurpose* buzzwords of an eco-friendly lifestyle, seek out unusual objects that will contain your herbs and reflect your personality. Secondhand and thrift stores, flea markets, salvage yards, garage sales, and hardware stores, among other places, offer endless possibilities.

And it's not just about repurposing worn out old pots. Vintage items, cast-offs, and other objects not typically used for growing plants make innovative choices. If it creates enough of a bowl shape to hold soil and plants, it could become a unique container. The (usually) lower price of such objects as compared to high-end pottery or purpose-made containers adds to the thrill of the hunt.

A metal tool box makes a fun home for scented geraniums.

This kind of container shopping is an adventure, so go with an open mind and be creative. Don't be afraid to improvise. Virtually anything that can withstand weather conditions, hold enough soil, and drain well can make a fine choice for growing herbs. If an object catches your eye, make sure it has a hole for drainage or evaporation or that you can drill or punch appropriate holes into it.

How a piece is made is more important than a faded finish. Don't pass up an object with an appealing shape or style that's past its prime if it will hold the weight and bulk of soil and plants. Evaluate whether you can revive a strong, sturdy container with a good scrub or a fresh coat of paint. Let the containers tell their stories as they become a part of the overall look and feel of your garden space.

OTHER CONTAINER CONSIDERATIONS

Choosing which of an endless variety of containers made with what materials is best for your garden begins with considering how each suits your needs and style. Here are some tips.

Budget
Well-made pots, whatever the size and material, are always a good investment. If you plan to grow a container herb garden for many years, spend the money on something that will last. Partner the design, style, and longevity of the containers with your budget.

Location
Plan your herb garden location before you start shopping for containers. Whether or not you use the elevator, carrying heavy planters up to a balcony or rooftop makes far less sense than opting for lightweight plastic. Hanging and vertical gardens can be good choices, but is the site capable of supporting them? Consider the cost, materials, tools, and skills it may require to make and hang them properly. In short, match the container to the location with care.

Keep in mind the size of the space where your container is going be placed to make sure it will actually fit. Measure the floor space, windowsill, window box, or vertical growing area.

Cultural conditions
The temperature of a space might also influence what type of pot to use. Natural materials that are well insulated stay cooler in sunny, hot areas. If the location is in partial sun, consider using planters made of stone or metal that gather heat during sunny periods and keep herbs warmer at night and on cooler days.

Quantity
The amount of available space and what types of herbs you want to grow determine how many pots you need. Balance your desire to grow all your must-have plants with being able to give them the space they need to grow abundantly. Large containers can accommodate several different varieties of herbs, smaller containers will hold only a single plant. If space limits you to a single, large pot, choose compatible plants rather than ones that will fight each other for space, water, and nutrients.

Pot shape

Considering how long you want a particular plant to stay in a particular pot can help you choose the best size and shape for a given container. If you anticipate moving your small herbs to larger containers or into the ground once the plants mature, pot shape is important. It's almost impossible to remove a plant from a narrow-necked container without breaking the pot or damaging the plant. Look for a wide rim and a narrower base. You should be able to gently tap the container bottom and easily pop out the plant.

By simple geometry of space, a square pot holds more soil volume and is therefore better for plants with fibrous or large roots systems than a narrow or cone-shaped container. If you plan to grow a woody or shrub-like herb with a large root system, consider a square-shaped container.

If you plan to mix different plants in one container, choose a large-diameter, open-rimmed pot that measures at least 24 inches (61 cm) across. Tall, narrow planters need to be heavy at the base to keep them from tipping over, especially when planting tall herbs and trees.

You might have the type of gardening space where you need multiple sizes of pots. I typically place herbs in 6-inch-deep (15 cm) terra cotta pots in my kitchen window during the winter. Measure small spaces, such as windowsills, countertops, or narrow ledges, to ensure that pottery on a drip tray or saucer will sit securely in the space. When the weather warms, I transition my window plants outside into larger pots. Outdoors, herbs should start out in containers a minimum of 12 inches (30 cm) deep. The larger the pot, the less it needs to be watered through the warmer months of the year.

Lemon balm will overfill a small pot quickly in one season. Plant it in a 10- to 12-inch (25 to 30 cm) pot and plan to repot after a few years to help keep it in check.

Pot depth

I am often asked how deep or wide a planter needs to be to grow herbs. A few things factor into the decision, primarily the type of plants you plan to grow and the planter's intended location. Herbs will certainly grow in a shallow pot, but more depth means more space to grow and a longer life in the pot. Herbs in a shallow pot may need to be transplanted or divided sooner.

To start out:
- Herbaceous plants, such as basil, calendula, and chives, need about 8 inches (20 cm) of soil depth.
- Herbs with long taproots, such as parsley and fennel, need a minimum of 15 inches (38 cm) of soil depth.
- Tree and shrub herbs need a minimum of 24 inches (61 cm) of soil depth.

Be careful to match large plants to the correct pot size. For example, you don't want a large, rounded plant, such as a chaste tree, in a small, narrow pot. Both a healthy root system and the overall scale and aesthetic of the design require a large pot. You want the plants and container to balance each other. A small container with big plants will appear in imminent danger of tipping over; a large or tall pot will distract from the look of small, creeping or tiny-leaved herbs, such as chamomile or thyme.

CACHEPOTS FOR CONTAINER GARDENS

Cachepot is a French term that describes a pot into which you place a smaller pot, most often containing a houseplant. These typically decorative, often colorful containers are a good way to disguise unattractive but perfectly good plant pots. Keep your plant in the ugly pot and place the whole thing inside a prettier cachepot. Because a classic cachepot lacks drainage holes, it prevents water stains on wooden decks, countertops, and similar surfaces. Empty the cachepot's accumulated water occasionally to avoid drowning the plant in the pot it contains.

CHAPTER 2

THE PERFECT PLACE FOR GROWING HERBS

Container gardening is all about the versatility to grow herbs wherever you have room: on a balcony, a patio, a rooftop, or the steps to your porch. It can be as simple as a planter outside the kitchen door or a hanging basket within reach to harvest a handful of mint, a windowsill filled with healing plants or a vintage wheelbarrow placed strategically to attract pollinators. You can position big or small container plantings to take advantage of the available space, make them easily accessible for harvest, and give the herbs their best growing conditions.

Where is *your* perfect place to grow herbs? Take a walk around and assess potential spaces for container gardens. Think about accessibility for watering, tending, trimming, and harvesting. There might be room to tuck something into an existing garden, or you could mount a window box on the house. Indoor herb gardens are particularly convenient to access. Floor space a premium? Try a living wall. As you consider all the possibilities, take into account what a container herb garden needs to grow well.

Herbs growing on a windowsill are simple to harvest.

LIGHT AND WATER NEEDS

The cultural needs and growing habits of the herbs you want to grow are critical considerations in deciding where to grow them. It rarely works to put sun-loving plants in a shady woodland garden or tall, dramatic plants, such as lovage, in a tiny space.

Most herbs need full sun (typically defined as a minimum of six hours) or as much bright light as you can give them. If direct sunlight is a challenge, look for areas that stay very bright for a good portion of the day—think quality over quantity. A western or southern exposure that gets at least four hours of afternoon sun might work, as may direct morning sunlight in an eastern exposure. The good thing is that, because you're gardening in containers, you can potentially shift plants around if a location isn't working.

Water needs are another consideration when scoping out a space to grow herbs. All plants need water, and containers especially need regular watering because the roots have no natural way to source ground moisture. Can you easily get a hose or watering can to the area? Water is sloppy, messy, and cumbersome; you want to make sure it doesn't become too much of a chore to get water to the plants.

If you place your containers near your water source, consider attaching a drip hose or drip system to a nearby hose bib or a zone on an existing sprinkler system. Drip systems apply water slowly, saturating the soil without disturbing the plants. You can turn it on and off manually, or add a timer to automate it.

HERBS FOR SHADE

Most herbs do best in full sun, but here are a few that will be fine in partial shade conditions:

alpine strawberries
bee balm
chives
cilantro
lemon balm
parsley
peppermint
spearmint
sweet woodruff
French tarragon
violas
violets

Purchase a decorative and stylish watering can to place near your container gardens. This part of the decor becomes a useful tool when it's time to water.

TAKING ADVANTAGE OF MICROCLIMATES

Herbs in containers can benefit from microclimates in the garden. A microclimate is just what it sounds like: a very small area, within a larger space, that has different climate conditions than the surrounding area. It typically means there is a difference from other parts of the garden in temperature, moisture, and other natural, weather-driven elements.

Sometimes a microclimate is very subtle and is created by a building, such as when the space under the eaves of a house stays dry because the roof edge blocks rain. This might be an excellent spot to grow drought-tolerant herbs. Stone walls that raise the air temperature a few degrees as they gather heat on a sunny day could be a spot to overwinter containers or warm them faster in the spring. Fences, rocks, water features, paved walkways, and patios all can have a part in creating a microclimate. Brick walls, asphalt paving, concrete, and other stone structures also gather heat and create warm balconies and decks.

To discover if you have a valuable microclimate for your containers, walk around and observe how the sun comes into your space. Where does it linger the longest, and how much of an area does it cover? There might be a tiny space where you can position a container garden to take advantage of just a few hours of very bright sun. Keep in mind that the sun's direction and angle will change throughout the year. A spot that's ideal in July might get far less sun in April. Ideally, make your observations several times during the year. And remember, even if you've already planted, your containers are movable. You can position them as needed to where your observations suggest they will thrive.

You can also create your own microclimates where they're needed. Most herbs won't fuss over too much sun, but if, for example, you're overwhelmed with watering, add a trellis to filter the day's most intense sunlight. A white or light-colored trellis in a shadier area can make things a little brighter by reflecting sunlight onto appropriately positioned containers, or paint a wall for a similar effect. Cluster pots together to shade each other and help hold moisture during dry spells and in hot climate areas. Move the containers away from each other for healthy air flow in wet or humid times of the year.

Take advantage of brick walls that gather heat from the sun by placing your planters there for warmth and protection from windy conditions.

PLACEMENT POSSIBILITIES FOR HERB CONTAINERS

To find areas where you can fit container herbs into your life, look up, down, and everywhere in between. Does your space allow you to hang a garden or place one vertically on a wall? Don't limit yourself to just placing containers at ground level. That is a good start, but explore your space and make the most of what you have.

Thai basil and hop-flowering ornamental oregano mingle with a ground cover oregano.

An oversized decorative jar with English lavender in a mixed planting creates a focal point in the midst of fennel, catmint, and other mixed perennials.

In existing garden beds

Whether your garden is large or small, you can mingle potted herbs with in-ground plantings to create an attractive display. You aren't limited to having just one spot for the formality of a set-aside herb garden.

Spaces that are already filled with plantings may have small gaps in which to set pottery, an especially good way to grow aggressive herbs. Placing interesting, artistic, or colorful planters is an easy way to add instant style to a boring space. Pottery is a more permanent touch of color that lasts when flowers fade. It is also a great design trick to fill in planting beds that have seasonal gaps.

Container gardens in the hardscape can serve other practical purposes. Lined up, they can create a hedge, a finite boundary, or definition in a planting border along a pathway. Filled with plants that attract butterflies or hummingbirds, containers provide food, shelter, and nesting space. Add your choice of culinary herbs in containers to a vegetable garden. Take a walk through your landscape to discover other ways to integrate herb-filled containers.

On balconies

Potted plants on a balcony can transform a harsh space into a garden oasis, especially in a high-rise urban setting with little or no living landscape otherwise in view. The lush window boxes on the balcony at my dentist's office give me a moment of calm as I sit in the chair and look out the window. Observe the view from inside your home. Visualize what you want your balcony garden to become. Work out a few logistics, and you can create a welcoming garden of fragrant herbs.

First things first: What's your balcony's weight load? Though it might seem like a balcony that can bear the weight of people using the space could also take the weight of a few planters, containers heavy with soil, plants, and water could tip the balance. If possible, check with a building contractor, landlord, or superintendent to know for certain how much weight your balcony can take and if there are any vulnerable areas. Lightweight containers are an excellent way to literally lighten things up.

Understand the unique cultural conditions offered by your balcony. Watch the angle of sunlight throughout the day, as it will affect how well plants grow. Wind swirling around a building can also be a factor. Avoid using tall, narrow pottery that can be unstable.

Every square inch of space on balconies is valuable. You can hang planters with basil and colorful annual herbs, such as violas and nasturtiums, on sunny railings. Tuck herbs that tolerate some shade into a living wall attached to the side of the building.

Hanging containers on a balcony railing simplifies harvests.

In entryways and on steps

Your home entryway or the back steps up to your house are opportune spots for containers. Pottery placed near doors can be welcoming. To keep these spaces visually

appealing, pay attention to details of how the plants look. Don't use herbs that you may continuously harvest. Use showy herbal flowers and leafy perennials to keep containers fresh and inviting. Add aromatic plants that release pleasant fragrances to greet guests as they brush past.

The size of your front steps and porch will dictate much about what you can grow there. You need containers with smaller bases. Look for narrow cone shapes that sit safely on steps but don't impede walking spaces and door access.

On rooftops

Whether atop a high-rise with a grand cityscape beyond or on a small fire escape to the top of a building, rooftop gardens are found space. Most roof spaces offer the abundance of light that gardens love, and containers are usually the only option for roof gardens. They invite birds and butterflies, grow food, and create a calm and nurturing haven above the bustle of a city. If a rooftop is your space to grow, let's climb the stairs and create a garden.

Gardening on a roof is similar to a balcony garden. Weight loads are an important thing to know. Check to see if there are any restrictions on weight and water use on a

Grow bags make a lightweight alternative for growing herbs on a roof.

flat roof. Consider how you will get soil, plants, and gardening tools to your space and where you will store them for everyday tending and care of your plants. Also consider safety before you get started. Check with your landlord or building superintendent for restrictions in using space on a fire escape or on other areas that would impede access in an emergency.

Rooftops tend to be exposed to sun and wind conditions that affect the choice of plants and containers. Tough woody herbs, such as rosemary, sage, and lavender, are better suited to windy conditions than delicate leafy herbs, such as basil. Metal containers can collect heat from the sun, asphalt shingles, and roof tiles. Use plastic pottery instead to help lessen the weight load and not overheat soil and plants. The wind can easily tip lightweight pots over, so use container styles with a wide base for stability. Top-dress soil in the pots with decorative rock or terra cotta shards to prevent loose soil from blowing out. The topdressing will also help keep moisture from evaporating too fast in heat and wind.

When planting tender leafy herbs on a rooftop or balcony, place them in sheltered locations out of wind.

On patios and decks

Container gardens add decor and flair to hardscaped outdoor living spaces. They lend greenery to areas where there is no inground garden space. They are also a quick way to upgrade a patio's ambience. Once you have the containers, plants, and

Fragrant heliotrope, variegated scented geranium, and rosemary mix for a touchable fragrant garden near a seating spot.

This fun little planter is hung on an otherwise boring concrete block wall.

soil, you can enhance your patio or deck in just a few hours. For example, add your favorite culinary herbs in a pot near the grill in an outdoor kitchen. In season, you will have fresh-cut flavor right at your fingertips. Add aromatherapy by placing abundant pottery filled with fragrant herbs near a seating space.

Position planters to make tending and watering convenient. Those placed directly on surfaces can stain and discolor wood or concrete. Pot feet, small pieces of pottery or wood placed under a container to elevate it slightly, prevent water from pooling beneath containers. For extra protection from water staining, place saucers under the pots or use cachepots (see page 29).

In living walls and hanging containers

Hanging plants and vertical planting walls are other ways to bring in herbs without taking up valuable floor space. This type of gardening is a unique way to get your herbs up close and within easy reach. Aromas drift at nose level, and you don't have to bend over to cut culinary herbs.

A vertical planting wall provides gardening space on a fence, the side of a building, or a balcony railing. Fabric pockets or flat-sided pots attach to the structure. Because most vertical or living herb walls are soil based, the structure needs to be able to support the weight of plants, soil, and water.

Hanging baskets are another opportunity to grow herbs when space is limited. They are an ideal way to bring height to a balcony or

Opposite: This vertical wall planter is a great way to grow a lot of herbs in a small amount of space.

patio with a roof overhang or eaves. Mount an herb basket where it can be well secured by a hanging bracket or hook. The basket should be far enough away from a building that if it sways in the wind, it will not risk being damaged or falling off the hook. The best herbs for use in hanging gardens are ones that tend to branch out horizontally or mound and tumble. If using upright plants, choose dwarf cultivars, center them in the middle of the planting, and use tumbling and spilling plants to soften the edges and grow over the side of the basket.

Living walls are easy to care for and can produce abundantly, with few pests or problems as long as you give the plants the sunlight, water, and nutrients they need. Lush, creeping herbs, such as specific cultivars of trailing plants, fill the pockets of wall gardens well to create a full, attractive garden. Larger wall pockets with a soil depth of at least 6 inches (15 cm) can hold dwarf upright herbs, such as garlic chives, basil, rosemary 'Blue Boy', French tarragon, curly parsley, and savory.

This large hanging basket made of wicker is filled with lush culinary herbs: rosemary, tricolor sage, and variegated thyme.

Plant a Hanging Herb Basket

SUPPLIES

wire basket, with hanger
coir basket liner
fresh sphagnum moss
potting soil

HERBS FOR HANGING BASKETS

For full sun:
climbing nasturtiums
dwarf catmint
prostrate rosemary
thyme
winter savory

For shade:
chamomile
pennyroyal
sweet woodruff
variegated mint
viola

Step 1: Stabilize the wire basket on top of a piece of pottery if needed to balance while planting it. Line the outer edge with green moss, then add a coir fiber basket liner to the inside. Fill with moist potting soil to just below the rim.

Step 2: Place trailing herbs around the rim to trail over the sides of the basket. Place taller, upright herbs in the center of the basket. Look for variegated and colorful leaved herbs to add contrast. Top the entire planter with soil and gently press down around the plants to remove any large air pockets.

Step 3: Secure the basket with a hanger. Water well and hang.

Living walls and hanging baskets do need some special attention to prevent crowding and underwatering. The plants are confined to smaller spaces and the soil tends to dry out quickly, so they need regular watering (sometimes twice a day in the heat of summer) or a drip system. Consistent grooming and watering will keep them lush and productive. Remove dead or diseased leaves and trim herbs to keep them compact and bushy. If plants become leggy and weak and start to drop leaves, you may need to adjust watering or thin and remove some plants to allow space.

In window boxes

A window box makes an ideal herb garden. It is accessible and changeable with the seasons as plants come and go. Place where the aromas of fragrant herbs can breeze through windows to create natural air fresheners on warm days. The narrow space of window boxes is best suited to slow-growing or compact plants that won't easily overtake each other. Tumble variegated thyme over the edges mixed with nasturtiums and prostrate rosemary. Tuck in calendula, violas, and saffron crocus to add seasonal colors. Basil, chives, curly parsley, and French tarragon are ideal for a kitchen window box of culinary herbs.

The windowsill and house siding where the box will be placed need to be strong enough to keep the box stable with brackets and take the weight of heavy soils and plants. If the window box is placed outside on a multistory building or deck, add extra secure fastening to avoid endangering people below. Classic window boxes are made of wood that weathers well and is less susceptible to rot, such as cedar, redwood, or oak. Metal racks with plastic inserts covered with moss or coco fiber mats are other good options.

Choose a type of window box that best suits the style of the house or building to which you are adding it. The length of a window box can be whatever your space allows, but the depth should be at least 8 to 10 inches (20 to 25 cm) to accommodate roots. Watering needs will be different for a window box, and you need to plan accordingly.

This metal window box decorates a potting shed window. Variegated apple mint and thyme blend with violas and leafy, colorful stalks of Swiss chard.

A collection of scented geraniums on a shelf drifts fragrance through an open window.

If the roofline extends over the box, rainfall will not be able to help you water. Plants in a window box also tend to dry out more quickly because of the airflow and close contact with warm building sidings or brick, so plan for water access or a drip system.

An alternative to a planted window box is a flat shelf built under a window on the outside. The advantage of a shelf as compared to a window box is that individual herbs in pots can be moved on and off the shelf. You can easily exchange plants in different seasons or add a freshly planted pot to fill an empty space as herbs fade or are harvested.

GROWING HERBS INDOORS

When the outdoor garden is at rest, bring herbs indoors and use them as houseplants. Mingling potted herbs into your home is another way to enjoy herbs throughout the year. Bring outdoor potted perennial herbs, such as chives and rosemary, inside before winter sets in. Start basil seeds to nourish a craving for basil in winter. Grow fresh starts of parsley for a fresh snip of flavor to garnish soup on a cold day. Brighten up a sunny window with potted herbs to enjoy any time, not just when they are in a seasonal garden outdoors.

Be sure your indoor herbs receive enough light or they'll grow long and lanky.

Creating the right growing conditions

There are limits to how well most herbs do inside the house, so take a few extra steps to create an indoor plant oasis for them. Consider what type of environment herbs grow in when they are outdoors. They prefer a sunny, warm spot with well-draining soil. A successful indoor herb garden will recreate those conditions.

Start by choosing a spot near a south-facing or west-facing window with as much direct light as possible. Put them on plant stands or in window boxes to get them up into the light. Even with a sunny window, your herbs will beg for more light. Keep them leafy and healthy by adding supplementary lighting. There are many grow light systems available.

Parsley, rosemary, sage, and thyme soak up warmth and sunlight in a west-facing window.

You can also create an attractive houseplant environment with decorative lamps. Replace a regular light bulb with an LED grow light to add red and blue colors that trigger healthy leaf growth. Mingle sun-loving herbs with other light-loving indoor houseplants, such as jade plant (*Crassula* spp.), snake plant (*Sansevieria trifasciata*), croton, areca palm (*Dypsis lutescens*), and hibiscus to create a vignette of lush greenery among your herbs.

Tips for Indoor Herb Gardening

- Place potted herbs in a south-facing window or another place that will get the best natural light. Rotate pots every few weeks to encourage even growth.
- Use grow lights if you don't have a place with a lot of natural light. Most spaces indoors will need auxiliary lighting to keep herbs from getting leggy and weak.
- Don't crowd multiple herbs into one pot. Plant one variety per pot to help with good air circulation and even light.
- Herbs grow best in air temperatures of 65°F to 70°F (18°C to 21°C). Monitor the air temperature of windowsills to make sure it is not too cold near the glass.
- Plant herbs in well-draining pots. Drip trays or saucers under the pots will catch water and protect household surfaces. Good drainage is vital to keep roots healthy, and the moisture collection in the drip tray will encourage humidity around the plants in the dry indoor air.

- Avoid overwatering. Check the soil moisture by poking your finger about an inch into the soil. If you can push in easily and it feels cool to the touch, no need to water. Check daily and water only when the top inch of soil is starting to dry out. Then, pamper herbs with warm water instead of a cold shower.
- Houseplants have no way to get nutrients for healthy growth unless you give it to them. Make sure to keep an eye on how the plants are doing. Typically, herbs require little extra fertilizer. You do not want to encourage a flush of fast growth, because the flavor and aroma of herbs are better when left a bit lean on nutrients. The ongoing watering of containers washes away nutrients at different rates in potted gardens. So simply look for changes to the plants; if herbs seem pale in color and weak, they could benefit from fertilizer. Use an organic liquid all-purpose fertilizer, mixed at half the strength. Do this once every two weeks until the plants perk up and look healthy again.
- To combat the drying effect of indoor heating systems, mist plants with fresh water from a spray bottle to provide humidity.
- Just as with your herbs outdoors, keep them trimmed and tidy to encourage leaf growth. Brush the dust off of leaves with a soft paintbrush.

Be diligent on pest watch. If herbs are brought indoors after spending the summer outside, inspect for bugs that may hitchhike their way into the house. Inspect upper and lower surfaces of leaves and along the stems. Give plants a good rinse to dislodge any bugs. Check the soil surface and around the edges of the root ball at the top of the pot. Gently turn the container on its side and check the bottom for slugs, ants, and earwigs, who like to hide in the dark recesses of drainage holes. Indoor herbs surrounded by other houseplants tend to attract spider mites, gnats, mealybugs, and aphids. Regularly inspect leaves both top and bottom and along stems, looking for white powdery or shiny surfaces that are unusual to the leaves. If you spot a problem, identify the pest first and find the best way to treat it. Sometimes taking the potted plant into a shower and gently rinsing it will dislodge webs and pests enough that it does not need to be treated further.

ORGANIC INDOOR SOAP

Give your plants and pottery a good cleaning and spot treatment with this easy, mild soap mix.

Use a mild organic liquid soap, such as castile soap. Mix ½ ounce (about 1 tablespoon or 15 ml) of soap to 32 ounces (4 cups or 940 ml) of clean water. Shake well and put the mix into a spray bottle. Shake liquid well before each use. Spray affected upper and lower leaf surfaces and allow spray to dry on the leaves.

Herbs that make good houseplants

Many herbs make wonderful houseplants. Here are a few of my favorites.

Aloe vera (*Aloe vera*): Easy to grow, aloe vera will live for many years indoors. The long, narrow, fleshy leaves contain a clear gel that heals burns and irritations when applied to the skin. The plant prefers filtered light and is fine near a bright window. A terra cotta planter with well-draining soil is best. Water the plant well but allow it to dry between watering. The demise of most indoor aloe vera is from rotting at the soil level caused by too much moisture around the base of the plant. As aloe matures, healthy plants will produce pups (baby plants) just off the parent plant. If the plants are in a larger pot, you can leave them as is for many years. Or, transplant the little babies to another pot when they reach about 3 inches (8 cm) tall.

Bay laurel (*Laurus nobilis*): This sturdy structural tree appreciates as much light as you can give it but is also not too fussy. Temperature swings inside will not affect it. It is very slow growing and does not need a large pot. Keep this in a container that you can take outside when the weather warms, and it will put on a burst of new growth over the summer. Bring it back indoors over the winter. Bay laurel tends to get scale and whitefly indoors, so keep an eye on it.

Lemon verbena (*Aloysia citriodora*): This fast-growing leafy plant smells like fresh lemons when the leaves are rubbed. It thrives in full sun and as a houseplant should be placed in a warm spot with a grow lamp for at least ten hours a day or in a conservatory space in full light. Place in a large pot, at least 15 inches (38 cm) wide and deep, so the plant can grow without being crowded. Good drainage is important, and regular trimming will encourage compact, leafy growth. Sometimes, if there is a dramatic temperature change, the plant will drop all its leaves. Don't despair; keep watering it every few weeks and new leaves will emerge from the woody stems.

Mint (*Mentha* spp.): Grow a mixed planter of mint indoors for fresh harvest all year. Mint likes an area that has bright light, but it does not need to be in full, direct sun. Mint is a fun houseplant for a bathroom that has a lot of bright, natural light. The plants appreciate the humidity from the shower and will add natural aromatherapy. Given the right indoor environment, mint will thrive with little care. Keep it regularly watered and in warm temperatures. Harvest leaves and pinch off flowers to keep it lush and full.

Scented geraniums (*Pelargonium* spp.) Scented-leaf geraniums are not the heavy flowering types most of us think of, that common red flower tumbling out of window boxes and hanging baskets in the summer. The leaves of these frost-sensitive, shrubby plants emit unique aromas, and collectors seek out the different varieties, often growing them indoors or in greenhouses to protect them from low temperatures. They will do best in a sunny location with extra light and warmth to keep them from getting leggy. They grow lush and abundant with extensive root systems and should be placed in their own pot. Regular trimming keeps them compact and full.

▶ **Stevia** (*Stevia rebaudiana*): Also known as the sweet leaf plant, stevia is native to Paraguay and likes warm, sunny growing conditions. The plant is sensitive to frost and will die in harsh winter areas. To help it thrive indoors, choose a sunny window and add extra lighting for robust, leafy plants. Keep the plant tidy by trimming off flowers and encouraging leafy growth.

Tea plant (*Camellia sinensis*): The true tea plant has dark green, glossy foliage. Indoors, it will appreciate a cooler spot (around 60°F [16°C]) with bright light. It does not need direct, hot sun. Humidity is necessary to help combat the effects of dry, heated air. Mist the plants daily or provide a humidifier if needed. Like all camellias, the tea plant prefers acidic soil and benefits by being moved outdoors in the summer, top-dressed with organic compost, and placed in a semi-shady area where it can get natural rainfall. Bring the plant back inside before the onset of cold weather.

COMMON AND BOTANICAL NAMES

The common name is what we generally call a plant in everyday usage; the botanical name is the precise Latin term for the plant, consisting of a genus and a species name. The botanical name is always in italics. For a few, such as aloe vera, the common and botanical names are the same, but for most plants they are different. For instance, bay laurel is *Laurus nobilis* in botanical parlance. Some plants have more than one common name, and one of those names might be from the botanical Latin. For instance, *Echinacea purpurea* is commonly referred to as both purple coneflower and echinacea.

In the world of herbs, it is typical to use common names, those that are most recognizable to most gardeners. So, why is it important to know the botanical name of an herb? Common names of many plants vary by geographic location, and some common names may even be attributed to two different herbs. If you're going to use an herb for medicinal or culinary purposes, making sure you're choosing the correct herb is important, to ensure both safety and the expected outcome.

In this book, I give botanical names in the detailed plant profiles but otherwise call herbs by their common names. Particular varieties of herbs (called cultivars, for "cultivated variety") may have their own special names, and those are indicated within single quotation marks: nasturtium 'Cherries Jubilee'.

Moving indoor herbs outside

If you plan to move herbs outside in the spring after wintering indoors, you need to acclimate them before leaving them outside all day and night. Much like you harden off new plant starts before they go into a vegetable garden, you need to reintroduce indoor herbs to outside conditions gradually. They are not used to wind, rain, and direct, hot sun.

Move potted herbs outside for a few hours each day when temperatures average in the upper 60s (about 19°C). Gradually increase their time outside every day by an hour. Cloudy days are best, but more important is that the temperature is warm and stable. You don't want to shock them too much with an abrupt temperature change. If the sun is direct or very hot, place the plants in a shady location for the first few days. Bring plants back in overnight.

Follow this routine for about a week. As they get used to accessing more light and having the wind on their leaves, they will start actively pushing new growth. They should be okay to leave out after that. Some tender plants, such as lemon verbena and basil, may take a few weeks. After they are acclimated, cut back any soft growth on perennial herbs to encourage new growth to be lush and leafy.

Opposite: Before moving herbs back outside in the spring, you'll need to acclimate them to outdoor conditions.

CHAPTER 3

CONTAINER GARDEN DESIGN

Herb gardens in containers, whether big or small, can create visual impact. They can accent a home, add color, or set a mood for a theme garden. You can place them where they are visible from a daily living space, such as a living room or dining room. Well-positioned window boxes add interest to the outside of the house and give those inside a premium view of birds and insects. Small, compact gardens in pottery can be a table centerpiece or a source of aromatherapy for those seated nearby. Living walls can serve as privacy screens or provide camouflage for ugly walls.

Think of your container as a garden in itself and consider how that garden looks year round as part of the décor where you set it. Maybe it is twenty containers to rim a large balcony or rooftop garden, or a single basil plant in a sunny kitchen window; it is all visual, and this is your chance to express creativity. Following are tips for using container color and style to your advantage, for composing pots that command attention, and for arranging pots in an eye-catching design.

A wooden window box filled with herbs, including tricolor sage and thyme mixed with annual petunias, is a cheerful focal point.

CONTAINER COLOR AND STYLE

Containers can enhance both the plantings in them and the surrounding area, blending in or standing out to make a statement. Glazed, embossed, elegant, textural, heavy or light, contemporary or traditional—the pot you choose is the style maker.

Connection with your home's style

If your home or gardening space has a distinctive design style, don't fight it. Containers will never mesh into the area if they clash with their surroundings. Metal and industrial materials overflowing with shimmering silver herbs, such as white sage, 'Silver Anouk' Spanish lavender, and 'Powis Castle' artemisia, create a contemporary look in an urban garden. Earthy, wooden planter boxes pick up on the strong natural elements typically used in a ranch or prairie style home.

Consider how a container can become part of the décor in much the same way you would select a picture for your wall or a lamp to complement your furniture. If a theme or mood calls for a real terra cotta pot, then use one; an imitation plastic one will never look right. If the color of the home or patio furniture will be in direct view of the pot, coordinate the effort. Remember, the colors of pottery are not seasonal; if it clashes now, it will always clash.

An elegant style maker, this classic pottery vase adds distinction to bronze fennel.

Pottery with tumbling garden geraniums and fragrant scented geraniums looks beautiful with a Victorian home.

The glaze, color, and shape of this pottery adds a contemporary feel to an outdoor patio. In this container garden, upright rosemary adds height as the 'Aurea' golden sage fills in the edges with color and texture.

Color accents

Color is a vital part of every garden design. Most herbs are varying shades of green, so the container takes on the extra duty of popping color and adding year-round interest. Well-thought-out touches of color added by garden pottery, furniture coverings, and throw pillows might just be the simple change you need to set a color mood for your container garden space.

Don't be afraid to match pottery color to your home or plant color to the pots. Interior and exterior decorators both use this design trick. The rhythm of color establishes unity and makes different pieces look in place with each other. Pick up a color from an existing element, such as a bold, burgundy-leaved basil added to a planter in an outdoor living space decorated with burgundy cushions on outdoor furniture. Set a glazed planter outdoors that matches your home color.

Another trick is to use repetition of one color. Pots in all one color make a small space feel larger. Too many pottery colors make a space feel cluttered. No need to be boring or overwhelming, just look at what colors you have to work with and go from there. Add an upscale look to a simple but beautiful herb by matching the color of the plant to the pot. Black Tower elderberry planted in a narrow, cone-shaped, black glazed pot makes a columnar statement; rosemary in a cobalt blue pot will take on an extra sparkle when the blue flowers bloom all up and down the branches of the plant.

Creeping rosemary, highlighted here with a blue glazed pot, blooms in the late winter.

Color play

Use your herb containers to create an illusion of space. Bright or vivid colors set in a distant corner on a deck or area in the garden will appear closer. The mass of color creates visual impact. If an existing space has intense, bold colors or the wall of buildings seen from a balcony seems imposing, soften the space with neutral gray, white, or black to play down dominant tones. Use a neutral wall or backdrop to set a scene with a bright or ornate piece of pottery.

Bright and bold mounding nasturtium 'Cherries Jubilee' tumbles over a window box mixed with thyme and rose-scented geraniums.

Shopping for bold-colored pottery.

Hot colors—reds, oranges, and yellows—bring an upbeat vibe. The colors of fire and the sun make a space feel warmer than it actually is. Create culinary gardens in bright red pots filled with bold herbs, such as cilantro, garlic chives, and Greek oregano, or add a feel of the south of France with mustard yellow pottery filled with bay laurel, chervil, fennel, and parsley.

Cool colors—shades of blue and green—soothe nerves, lower blood pressure, and create a sense of relaxation. Green makes a space feel cooler. Use shades of blue or green containers for natural spa and healing herb gardens to bring a calming atmosphere.

Jewel tones—deep violets, purple, burgundy, and royal blue—add luxury and richness. Plant a deep purple or burgundy container with leafy herbs, such as purple shiso, contrasted with the soft, velvety leaves of purple sage and Thai basils. Use jewel-tone pots with herb plantings that are all shades of green to add color to a space.

COMPOSING POTS

Flowers, leaves, texture, drama; your herbal container can have it all. Mix, match, contrast, and experiment to create attractive combinations. It does take a certain amount of vision to know what the result will be as the plants mature, but trying different groupings is the way to find out. In your container design, take into consideration annuals, perennials, and biennials to design for all seasons.

Look for combinations that allow each plant to have individuality. If your chosen herbs are all shades of green, pay attention to the shape and size of their foliage and how they sit next to each other in mixed containers. Typically, plants that are standouts and bold by themselves are good in mixed containers. Place plants with bold leaves next to delicate ones. Plants that are too similar in leaf shapes or sizes tend to blur together, and the container has no interest or dimension. Plants that are aggressive and fast growing should be planted by themselves or in pots large enough for every plant to have room.

Make a Portable Pot Template

Mixing and matching plants can be complicated enough, but how does it all fit inside a container? To make it easy, create a template of your container to take to a garden store or use it in the garden to try out how plants look next to each other. This is a good tool to visualize how plants look side by side and in the shape of your container.

Lay the poster board or cardboard on a flat surface. If your pot is light enough, overturn it onto the paper and trace around the rim (be careful not to permanently mark the pot). Remove the pot. For heavier pots, place the poster board or cardboard on top and trace the same area from underneath. Cut around the outside edge of the outline you made. You now have a template of your pot.

Take this to the garden center and place it on a flat surface. Try setting different herbs on your template as you would plant them in the actual pot. Step back and look at the color and texture contrasts. Add plants to fill in and arrange until you like what you see. Don't be afraid to repeat plants all around to balance the design. Check your selections to make sure they will grow well together and have the same sun, shade, and water needs.

> SUPPLIES
>
> stiff poster board or cardboard, slightly larger than your planting surface
> black marker
> measuring tape
> heavy-duty scissors

Opposite: Curly parsley adds an herbal touch in a mixed ornamental planting. The hues of green and gold contrast with the deep-jewel-toned pot.

General design tips for plantings

The mixture of plants in a single container or the grouping of individually potted plants can make or break a good container garden design. Here are some things to keep in mind.

For mixed plantings:
- Group together plants that have the same cultural needs for light and water.
- Place taller-growing herbs, such as dill and fennel, toward the back of a container or at the center, where they won't shade out low-growing herbs.
- It's okay to plant herbs close together to create an instantly full pot. Based on what you put in the pot and its purpose, you may be harvesting things before the end of the season anyway.
- Pay attention to specific cultivars. Colorful variegated versions of herbs can bring a container garden to life. Dwarf forms are especially helpful in containers to help keep appropriate size and shape.
- Cilantro, calendula, basil, parsley, and stevia are all excellent herbs to tuck into mixes. They add nice leafy texture without overtaking other plants.

Bright blue borage grows through a rose-scented geranium.

Use leafy cilantro to fill in around woody plants in mixed herb garden containers.

This lemon verbena has room to grow large in its own pot.

For individual herbs in pots:
- Balance the plant's mature height with the height of the pot. A good rule of thumb is to match them.
- Create individual containers of the same herb for lushly planted bouquets to tuck around other, larger pots. In the spring, evenly space three to five small starts or cuttings in a 12-inch (30 cm) pot. Calendula, basil, nasturtiums, borage, violas, and cilantro all do well with multiples in the same pot.
- Single perennial herbs, such as lemon verbena, sage, lavender, and rosemary, should start out in a minimum 12-inch (30 cm) pot and be allowed to fill in over a season.

Flowering herbs for seasonal color

In container gardening, flowers can refresh a container garden's appeal in season. I'm a big fan of matching flower colors to pots for an added designer moment at bloom time. The container never fades, but flowering herbs add lush seasonal color.

Red/purple/blue blooms:
anise hyssop 'Blue Boa',
   'Blue Fortune', 'Blue Giant'
bee balm
borage
catmint
chives
lavender
oregano 'Herrenhausen',
   'Hopley's Purple'
purple coneflower
violas

Yellow/orange blooms:
anise hyssop 'Acapulco',
   'Poquito', 'Shades of Orange'
calendula
lady's mantle
marigold 'Lemon Gem'

White or pastel blooms:
chamomile
echinacea 'White Swan'
elderflower
garlic chives
oregano (Greek, hop flower)
roses
sweet woodruff

Herbs with colorful foliage

The foliage of herbs adds real drama to container gardens. Color and shape are the best way to make combination plantings visually exciting. A single herb with beautiful foliage in an ornate pot makes a statement.

Silver and blue foliage:
artemisia 'Powis Castle'
curry plant
lavender
rue 'Jackman's Blue'
sage
santolina
thyme 'Silver Posie'

Bronze and purple foliage:
basil 'Dark Opal',
    'Purple Ruffles', 'Amethyst'
bronze fennel
elderberry Black Beauty®, Black
    Lace®, Black Tower
sage 'Purpurascens'

Gold foliage:
anise hyssop 'Golden Jubilee'
oregano 'Aureum'
sage 'Icterina'
santolina 'Lemon Fizz'
thyme 'Archer's Gold',
    'Doone Valley'

Variegated/color-splashed foliage:
lavender 'Meerlo', 'Platinum Blonde'
pineapple mint 'Variegata'
sage 'Tricolor'
scented geranium
    'Lady Plymouth',
    'Chocolate Mint'

Thrillers, spillers, and fillers

A well-known formula for container garden design is to use "thrillers, spillers, and fillers." Thrillers provide height and drama, spillers cascade from the edges, and fillers cover the space in between.

Cardoon (ornamental artichoke) is the thriller in this mix, next to a filler of chives.

These herbs are thrillers, tall attention getters for the middle or back of a container that create mood and attitude:

angelica
cardoon
elderberry
fennel
globe artichoke
lovage
red castor bean

The bright green leaves of sweet woodruff, sprinkled with white blooms, along with chamomile soften the edges of this potted purple duo.

These "spiller" plants tease the sides of the pot and tumble over to soften the edges:

chamomile
oregano (creeping golden)
oregano 'Kent Beauty' (hop-flowering
   ornamental varieties)
rosemary 'Huntington Carpet', 'Irene'
   (trailing varieties)
sweet woodruff
thyme 'Pink Chintz' (creeping varieties)

These plants grow enough to fill in the gaps between the tall, dramatic plants and the creeping plants that spill over the sides:

basil
germander
lavender
mint
parsley
rosemary 'Arp', 'Roman Beauty' (upright varieties)
rue
santolina
thyme (upright varieties, such as variegated lemon)

'Jackman's Blue' rue.

GROUPING AND ARRANGING POTS

Once you have composed your herb containers, give careful thought to how you group and arrange them. You are adding architecture to an area. Frame doorways or entry points to a garden. Create a patio edge or define a walkway with a line of multiple pots in the exact same style, size, and color, each pot filled with a different herb.

Oversized terra cotta pots filled with mixed herbs arranged along a pathway create a lush potted garden display. Different styles of pottery add interest, while the color match of terra cotta creates continuity all along the walk.

An upright rosemary in a classic amphora-style pot tucks around a corner to create a tall, narrow focal point.

Compose a scene by how you arrange pots. Create different tiers and multiple levels of interest by turning empty pots over and setting planted ones on top. Use darker and lighter shades of the same color pot when using tiers or elevating some pots above others. The color shades will give interesting depth to a grouping.

Establish a vista or focal point with a well-positioned container. Add large herb trees, such as bay laurel or chaste tree, in containers as a backdrop to other potted herbs or as a way to direct the eye to follow along a building or path.

Using herbal trees and shrubs for architecture

Tree herbs, such as the healing herb witch hazel, can grow to 8 feet (2.4 m) tall in a garden setting. Starting them in a container garden tends to keep their roots and growth habit tidier so they do not spread as they would directly in the ground, but the plants will be healthier if their roots don't get bound up. They may stay more compact in height as the containers keep them in check, but they still need enough space to accommodate a healthy root system.

Herbal trees and shrubs, such as those listed below, can serve an important architectural purpose in the design of your container garden. The great advantage of growing them in pots is that you can move them around and place them in different groupings and configurations to achieve the best visual effect as your garden goes through seasonal changes.

Bay laurel (*Laurus nobilis*): This large, shrubby evergreen tree is popular trimmed and shaped into a pyramid, cone, or ball-style topiary. The leathery leaves have a distinct earthy aroma and flavor and are used to season soups and stews.

Carolina allspice (*Calycanthus* spp.): A sizeable leafy shrub that gets its common name from the spicy, aromatic flowers, it is not the traditional woody spice called allspice (*Pimenta dioica*). Place near a patio or seating area so you can enjoy the intensely fragrant flowers when they bloom in early summer. Carolina allspice has multiple documented uses as a medicinal herb. It's popular in its native Appalachian region of the United States. In container gardens, it is a tall, striking ornamental plant. Pot it up by itself in a container to allow it to grow large and flower freely.

Calycanthus raulstonii 'Hartlage Wine' in bloom.

Chaste tree produces spires of beautiful blooms.

Chaste tree (*Vitex* spp.): This large, rounded, deciduous flowering shrub does best in soil that is kept consistently moist. It is important that the pot drains well so that the roots don't become waterlogged. Chaste tree can be trained as a lovely multi-stemmed accent tree in a large container. Abundant, large, lilac-colored blooms flower freely in late summer and attract hummingbirds and butterflies.

Elderberry (*Sambucus* spp.): Elderberry is native to large areas of North America. The berries are popular as a cold remedy and as an immune system booster. Ornamental varieties of elderberry with rich burgundy or bright golden leaves fill large container gardens for dramatic, colorful height. Black Tower (*S. nigra* 'EIFFEL 1') and Golden Tower (*S. nigra* 'Jandeboer001') are columnar-growing varieties that add a tall accent without taking up too much space in containers.

Ginkgo (*Ginkgo biloba*): A slow-growing tree, most varieties of ginkgo (also called maidenhair tree) become large and are popular as a dramatic and long-lived landscape tree. Bushy and compact varieties, such as 'Chi-Chi' and 'Mariken', are suitable for containers. The leaf texture and shape make these perfect for architecture in a large mixed herb container.

Tea plant (*Camellia sinensis*): A glossy-leaved, broadleaf evergreen covered with small creamy white flowers that bloom in late fall and early winter, its leaves are harvested and processed for tea. This camellia does well in a large container garden. Use it as an evergreen screen or backdrop to a smaller grouping of containers.

Witch hazel (*Hamamelis* spp.): A medium-sized deciduous tree with an interesting horizontal branching habit, witch hazel gives an attractive silhouette to a container garden. Small tufts of fragrant flowers add late winter interest. Plant in a large container and surround the base with low, mounding understory herbs, such as chamomile and sweet woodruff.

Opposite: Two favorite immune system boosting medicinal herbs: elderberry with dark purple berries and purple coneflower (*Echinacea purpurea*).

Grouping deer-resistant herbs for protection

Sometimes design needs to take a practical approach while still being stylish. In gardens where deer are a challenge, containers are not immune from browsing if the deer can reach them. The good news is that many herbs are deer resistant, and pots of these herbs can be grouped and arranged to protect other plants that deer consider yummy.

Fill large containers with plants that deer do not like and strategically place them in the garden as a barrier to those they find tasty. Create a pottery hedge that stops deer from going through an area or line a pathway to funnel or guide deer through a space and away from plants that are browsed.

TOP DEER-RESISTANT HERBS

anise hyssop
calendula
catmint
chives
hyssop
lavender
lemon balm
mint
rosemary
rue
sage (common garden)
French tarragon
thyme

Lush, low potted bowls of 'Berggarten' sage surround a 'Gertrude Jekyll' rose to keep deer from eating the blooms.

PART 2

GROW THIS! EASY HERB CONTAINER PROJECTS

Choosing herbs to grow based on a theme is an exciting and fun way to learn more about this diverse category of plants. The following chapters offer easy recipes for rewarding container gardens planned around themes, such as culinary herbs for kitchen gardens and medicinal herbs for healing gardens. Whatever your journey in container gardening, use these recipes to begin creating your own mixes. Let's get creative!

CHAPTER 4

CULINARY DELIGHTS

FAVORITE HERBS FOR COOKING AND FOR MIXING AND MINGLING WITH VEGGIES IN CONTAINERS

Packaged and processed herbs will never compare to the garden-fresh ones you grow. Adding culinary herbs to container gardens gives you flavor with no limits. Love basil? Grow multiple pots and sow successive crops to have this aromatic, colorful annual throughout the warm growing season. Place a pot near the kitchen door to quickly access fresh leaves as the pasta sauce simmers on the stove.

Culinary herbs also add beauty to containers. Variegated mint in a large pot in the midst of a culinary herb garden will add color throughout the season as edibles are harvested. The texture of chives and parsley will stand out when combined with other culinary herbs in containers.

Many of the herbs photographed and featured in this section are encouraged by cutting. The more you snip, the more you stimulate new growth. More growth, more herbal flavor, more to cook with!

HARVESTING FRESH HERBS FOR COOKING

A culinary garden is intended to use for garden-fresh flavor all growing season. To keep containers looking nice while you enjoy the herbs in cooking, harvest but don't defoliate your plants. Trim leaves and stems around and under bushy plants or select stems that don't affect the plant's overall look. Here are some general rules to keep plants producing well.

Annual herbs: Leave at least 5 inches (13 cm) of leafy growth and remove flowers to keep leaf production. Basil is the exception; if you are making a batch of pesto, you need a lot of leaves. Add new basil plants back into the space left behind after a basil harvest, if needed.

Perennial herbs: Leave 4 to 6 inches (10 to 15 cm) or at least a third of the plant to continue growth. Shape the plant to look natural as you cut.

Most herb lovers and foodies will recognize these herbs. Traditional and easy to grow, these are the best flavor makers for seasoning food. Some herbs in this mix will stay evergreen in mild winters or are okay with a light frost. The hardiness of these herbs gives an extended season of harvest and container garden beauty. Best in full sun, set this grouping where you can harvest from it easily throughout the season. Place them in edible gardens around raised beds as design accents or set up right near a door close to the kitchen to give quick access when cooking.

> *"Start in the garden. Grow your favorite herbs.*
> *Grow them because you like the flavor,*
> *then let your taste buds be your guide."*
> —A TASTE FOR HERBS

This project uses chimney flues as planters. Clay flues are manufactured as a liner for use in chimneys to direct smoke and other materials away from the building and out into the air. They are heavy, well made, and relatively inexpensive. They come in various shapes and sizes. Open on both ends, chimney flues can be set directly on the ground in a garden space to work like a small raised bed. Simply fill them with soil and plant away. Flues also make a great barrier to contain aggressive herbs, such as mint and oregano; planted individually, they will flourish without taking over. Stack chimney flues at different levels to create interest and accessibility.

Flues can be found at masonry or building supply stores. Or, check out antique shops and building salvage stores for vintage flues; older ones can be wonderfully decorative. A note of caution:

It is not easy to know the history of an old flue, and there may be remnants of chemicals you don't want to contaminate your herbs. Use them as a type of cachepot. Slide a plain pot down inside the flue for an easy decorative alternative to planting in them directly.

Herbal profiles

Basil 'Red Rubin' (*Ocimum basilicum*): Warm season annual, tender to frost. Easily grown from seed, basils are a must for every cook's garden. The leaves of 'Red Rubin' basil stay a deep burgundy color through the growing season. A well-behaved plant, this cultivar is easy

---

PLANTS NEEDED
basil 'Red Rubin'
chives
dill 'Bouquet'
oregano (Greek)
parsley (flat-leaf)
peppermint
savory (winter)
shiso (purple)
thyme (lemon)

to keep compact, making it a good companion with other plants in containers. This beauty does double duty in a container by adding deep, vibrant leaf color plus the classic Italian basil scent. Grow basil in a warm, sunny place and keep flowers pinched for best leaf production. Start successive crops of basil seed over a few weeks to have a good supply to tuck in open spaces of containers throughout the garden. Other burgundy-leaf basil varieties to look for include 'Dark Opal', 'Amethyst Improved', and 'Purple Ruffles'.

Chives (*Allium schoenoprasum*): Hardy perennial. Easy to grow from seed and very abundant. All parts of the plant are edible, including the fluffy pink flowers. To keep vigorous stem production and nice grass-like texture in the pot, deadhead the faded flowers (or add them to a salad while they're still fresh!). The plants have fibrous roots and appreciate a deep container but will not overtake plants around them. Divide the clumps every few years for a long-lived container garden of chives. To keep the plant attractive while harvesting, cut the older outer stems for use and leave the smaller new stems in the center of the clump. Garlicky and flavorful, chives are a non-fussy herb and a mainstay of every culinary container garden.

Dill 'Bouquet' (*Anethum graveolens*): Annual. Easy to start from seed. Plant seed directly outside in pots in the spring after danger of heavy frost. Dill has a long tap-

root and does best in containers that are at least 12 inches (30 cm) deep. Other varieties that stay compact and attractive in containers include 'Dukat' and 'Fernleaf'. The tall, ferny foliage adds height and texture to the back or middle of a container. Cut off new flower shoots during the peak growing season to keep production of the aromatic leaves. The tiny leaves, which are best used fresh, add a touch of licorice flavor to salads, roasted vegetables, and baked seafood dishes. Later in the growing season, allow the plants to flower. The leaves will not be flavorful but the aromatic flower heads are attractive. Use the heads and seed to flavor pickled cucumbers, beans, and other vegetables.

Oregano, Greek (*Origanum vulgare* ssp. *hirtum*): Perennial. Fast growing and likes to be in the warm sun. Greek oregano varieties, such as 'Hot and Spicy' and 'Kaliteri', are among the best types for a culinary garden. They have a strong, spicy flavor that holds up well under the heat of cooking. One way to recognize Greek oregano is that the flowers are white rather than the purple of common oregano (*Origanum vulgare*). Keep the flowers lightly sheared off through the growing season to keep plants bushy. The young, tender leaves have the best flavor. Oregano is best planted by itself, or its aggressive roots may overtake everything else in the pot.

▾ **Parsley, flat-leaf** (*Petroselinum crispum* var. *neapolitanum*): Biennial, typically treated as an annual. Parsley will produce an abundance of leaves its first season, and then it flowers and goes to seed in the second year of growth. The seed can be slow to germinate, so it can be more efficient to buy fresh plants every year. In containers, parsley's deep green leaves contrast well with other herbs; it adds a real wow factor when planted next to burgundy leaf basils. Also known as Italian or plain parsley, the flat-leaf varieties have a deeper, richer flavor than the decorative curly parsley (*Petroselinum crispum*), making it a better choice for the culinary garden.

Peppermint (*Mentha* x *piperita*): Perennial. Grow mint in its own container to keep the plant's assertive nature in check. The plant goes dormant and dies back to the ground in the winter, although it is not unusual for mint to stay evergreen in mild climates. Regularly trim back old growth and flowers throughout the season to keep fresh growth of young leaves, which have the most robust flavor. Divide the plants every two to three years so they do not become root-bound. Mint is a versatile culinary herb that can go savory or sweet. The flavor can add sweetness to tea, jelly, and desserts but also power up a savory dish when mixed with other pungent flavors, such as pepper or garlic.

Savory, winter (*Satureja montana*): Perennial. A low shrubby plant that fills in open spaces well in containers. A mass of small white flowers covers the plant in late summer. Shear faded flowers to encourage new growth. Leaves are spicy and reminiscent of a mix of other pungent Mediterranean herbs. Use savory as a substitute in cooking if you don't have thyme, rosemary, or oregano available. Another type, summer savory (*Satureja hortensis*), an annual grown easily from seed, makes a delicate, leafy filler in container gardens. Summer savory has a milder, slightly sweeter flavor than its winter counterpart. Either can be used in recipes interchangeably.

Shiso, purple (*Perilla frutescens*): Annual. Large and leafy, shiso in containers is like the coleus of the herb world. The deep-burgundy variety adds dramatic ornamental interest. Another good decorative variety to look for is 'Britton', which has green leaves with deep red undersides. Shiso stays full and lush all summer if you snip off the flowers. It is an excellent tall herb to use in the background of a planter. Its dark color gives other, smaller-leaved plants a chance to stand out in the design. The leaves have a distinctive aroma reminiscent of cinnamon, cloves, and anise, popular for use in Japanese, Thai, and Korean cooking. The leaves of the purple varieties can also be used to add color to vinegar and soups.

▸ **Thyme, lemon** (*Thymus* x *citriodorus* 'Variegata'): Perennial. This tough, bushy plant makes a great filler in a container. It also makes a nice companion in mixed herb planters. It stays demure and will not overtake space, while the golden color highlights plants growing next to it. Leaves may stay evergreen in mild winters; otherwise, this deciduous

plant will come back better year after year. In the spring, as new leaves emerge, they tend to be very attractive to slugs, so keep watch and take action if needed. This lemon-scented cultivar is a favorite in cooking. It stays true to its rich thyme flavor with an aroma of lemon that seasons fresh foods, such as salads and vegetables. Sprinkle on baked chicken or fish just as it is finishing to impart the delicate lemony flavor through the meat.

OTHER TRADITIONAL CULINARY HERBS FOR CONTAINER GARDENS
bay laurel
cilantro
French tarragon
rosemary
sage
summer savory

This small batch of richly flavorful Italian herbs in a garden is convenient to keep near outdoor cooking spaces. Harvest and use fresh as the dish is cooking or marinate their pungent flavors into meats and sauces. Create a tabletop garden or centerpiece for your outdoor dining table. Keep a small pair of scissors nearby so dining guests can snip fresh herbs to sprinkle over vegetables in the *contorno* course or on *insalata*.

This little potted garden uses individual glazed pots inside an attractive wicker basket, perfect for on the counter of an outdoor kitchen and easy care for harvesting all summer. When the perennial plants outgrow these small pots (probably within the first year), plant them in larger containers or out in the garden. Complete your *delizioso* flavor journey with these compact edibles planted in other containers nearby: 'Little Napoli' patio Roma tomato, 'Lizzano' cherry tomato, 'Little Prince' eggplant, 'Lacinato' kale, and radicchio.

PLANTS NEEDED

basil 'Genovese'
oregano 'Hot and Spicy'
parsley (Italian flat-leaf)
rosemary 'Barbeque'

"'Sei sempre in mezzo come il prezzemolo!'
Its literal translation is, 'You are always in the way like parsley!'
It deals with the common use of parsley in Italian recipes.
It is a euphemism to say that you mingle in things that are
not your own; you are always in the way."

—CHARMINGITALY.COM

Herbal profiles

Basil 'Genovese' (*Ocimum basilicum* 'Genovese'): Annual. Easy to start from seed. One tip for the success of basil in containers is to place them outside only after nighttime temperatures are consistently above 45°F (7°C) and the warmth of summer has arrived. Basil will fuss in cooler temperatures and shut down leaf production. Basil (*basilico* in Italian) is the one herb that it seems you can never have enough of. Keep up production by sowing a new crop to add to containers as you harvest mature plants from them. 'Genovese' is one of the more intense and true Italian varieties of basil. Others to look for include 'Dolce Fresca', 'Italian Pesto', and 'Profumo di Genova'.

Oregano 'Hot and Spicy' (*Origanum vulgare* 'Hot and Spicy'): Perennial. Small, soft leaves are loaded with aroma. This oregano has an irregular growth habit and will tumble over the side of a pot rather than stay upright like most oregano. Trim stems as needed to keep a tidy shape. The taste is much as its name implies: It has a peppery bite that is strong and pungent in cooking. The aroma of oregano reminds me of my sweet Italian grandma. After a sauté of garlic cloves in olive oil that made the kitchen smell amazing, she would simmer the ingredients of a red sauce for hours. The oregano (*oregano* in Italian) aroma would permeate the air, with basil added at the end for a mouthwatering sauce ready to pour on meats or pasta.

Parsley, Italian or flat-leaf (*Petroselinum crispum* var. *neapolitanum*): Biennial, typically treated as an annual. The lacy leaves of flat parsley are as decorative in a container as they are flavorful in cooking. The deep green leaves will grow abundant in the warm season and persevere through the year in mild weather. Parsley is a nice no-nonsense addition to container gardens. It adds great texture with little fuss and keeps within its space in mixed containers. In a small pot by itself, parsley will be fine for a few seasons. Due to its biennial nature, it will produce flowers and go to seed the second year of growth, so you will be replacing it anyway. Parsley (*prezzemolo* in Italian) is a common ingredient in Italian dishes to offset extra-spicy foods. Condiments made from parsley, such as gremolata, give the tongue a chance to rest from spiciness and be able to enjoy the flavors of the complete dish. Bonus use: Parsley is a breath freshener, so make sure to add a fresh cutting as a garnish to chew on after a garlicky meal.

Rosemary 'Barbeque' (*Rosmarinus officinalis* 'Barbeque'): Perennial. Rosemary is an evergreen shrub in areas with mild winters. It is an excellent upright, tall addition in potted gardens. It likes to have some root space, so it will only be useful in a small pot for one season. Transplant rosemary to a larger container after enjoying it in your small tabletop garden over the summer. I have been asked many times why rosemary sometimes spontaneously dies in potted gardens. Most of the time, the problem is at the soil level and below. Rosemary can become root-bound quickly and not get an even distribution of water for the plant to uptake. The soil always seems dry in root-bound plants, so our nature is to keep watering them. Our misplaced kindness can cause all those bound-up roots to rot. Keep an eye on your plant and transplant it to a larger pot before it gets root-bound. Rosemary (*rosmarino* in Italian) is a pine-like, pungent herb that complements other heavy spices, meats, and vegetables used in Italian dishes.

> **OTHER FLAVORFUL HERBS FOR ITALIAN CULINARY GARDENS**
> bay laurel (*alloro*)
> fennel (*finocchio*)
> marjoram (*maggiorana*)
> sage (*salvia*)

IS IT ROSEMARY OR IS IT SALVIA?

If you are already an herb aficionado, you know that Salvias are a big plant family, and in the herb world they belong to the common garden sage group. In February 2017, a team of researchers from the United States, Mexico, and Asia issued a report in the journal *TAXON* stating that recent DNA samples of rosemary, the genus *Rosmarinus* showed that it should be listed in the Salvia family. The botanical name

Salvia officinalis is already used for common garden sage, so it could not also be given to the grouping of *Rosmarinus officinalis*. The new classification for rosemary has become *Salvia rosmarinus*. Home gardeners are hard-pressed to keep up with the botanical naming of plants and their DNA discoveries, so in this book, and probably forever in my mind, it will still be *Rosmarinus* or, as we fondly call it, rosemary.

BROILED HERB AND CHEESE TOMATOES

4 fresh tomatoes, halved

2 or 3 fresh basil leaves, stems removed

1 teaspoon fresh oregano leaves

1 tablespoon (4g) flat-leaf parsley

½ cup (45 g) fresh, coarsely grated Parmesan cheese

salt and pepper to taste

Preheat your broiler. Place the tomato slices on a baking sheet, cut side up. Chop the basil, oregano, and parsley together until well mixed, then sprinkle evenly over the tomatoes. Top with fresh Parmesan cheese and add salt and pepper to taste. Broil approximately 3 to 4 minutes until the cheese begins to melt and lightly brown.

The herbs used in French cooking are part of the secret to the food's colors and distinctive flavors. Most are from the Mediterranean, and these richly aromatic plants make beautiful container gardens. Add touches of—or exuberant—inspiration from French gardens to give *your* garden a Francophile flair. The glazed urn pictured opposite mimics the sun-washed hues of the South of France and is a natural play on French country garden style. Corrugated zinc planters or metal urns with touches of patinated copper stylize the look of a chic Parisian courtyard planting. Plant dwarf annual sunflowers and cobalt blue cornflowers, also known as bachelor buttons, to give seasonal color. Tuck in compact-growing, container-friendly vegetables to make mini potagers.

PLANTS NEEDED

Emerald Wave® bay laurel
marjoram
rosemary 'Blue Spires'
sage 'Variegated Berggarten'
French tarragon
thyme (French)

Herbal profiles

Emerald Wave bay laurel (*Laurus nobilis* 'Monem'): Woody shrub or small tree. 'Monem' is a form of common bay laurel with long, narrow leaves that have a bit of a wavy texture along the edges, making this an interesting plant to show off in containers. A fragrant culinary plant and a nice tree for container gardens, its habit is narrower than the regular bay but just as slow growing, making it an excellent plant to shape as topiary. Leaves of the bay (*laurier* in French, also known as sweet bay) are popular for use in soups and stews because their strong, pungent flavor holds well in long-simmering culinary dishes. The sturdy, shiny green leaves can be picked and used any time. Typically, you need only one or two leaves to season a recipe, so your container garden tree will always look leafy and not overharvested.

Marjoram (*Origanum majorana*): Tender perennial, typically grown as an annual in colder climates. This plant will stay handsome and shrubby in a pot and can be used to fill in gaps around taller plants, such as rosemary and bay. The tiny white flowers look like little pearls along the stems. I tend to let the plant flower freely over the season for the lacy look it adds to the container. Once the flowers brown and look unsightly, shear

back the flower stems to the central, leafy part of the plant. It will usually flush out new growth and try to flower again at the end of the summer. Often confused with its close relative oregano, marjoram (*marjolaine* in French) has a sweeter, more subtle flavor. Because of its delicate nature, add marjoram to dishes during the last five to ten minutes of cooking to keep a strong flavor. Best used fresh because the sweet taste does not hold well through the drying process.

Rosemary 'Blue Spires' (*Rosmarinus officinalis* 'Blue Spires'): Perennial. Tender perennial in areas subject to heavy freezing temperatures. The name of this rosemary describes its lovely winter flowers, reminiscent of the deep blue skies of Tuscany. Upright and easy to keep from sprawling as some rosemary can do, this is an attractive, silvery-foliage cultivar to put in containers. It adds height and a linear shape and plays well against more voluminous herbs, such as sage and thyme. Earthy and pungent, rosemary (*romarin* in French) is an abundant herb that works as well on roasted vegetables and meats as it does in soups and stews.

Sage 'Berggarten Variegated' (*Salvia officinalis* 'Berggarten Variegated'): Perennial. The variegated leaves of this sage make it a real standout in container gardens. True to the silver color of regular 'Berggarten' sage, the accent of creamy, irregular variegation makes this one of my favorites to add a mid-layer of volume to a planter. In mild climates, it can remain evergreen, and it's easy to keep compact and full by trimming off any stems that get woody and out of bounds. In cooking, this variety of sage (*sauge* in French) is somewhat milder than common sage and not at all bitter.

Tarragon, French (*Artemisia dracunculus* var. *sativa*): Perennial. This is the true French tarragon and does not propagate by seeds. Look for starts in local garden stores to get the unique licorice-flavored leaf. Don't be fooled by other tarragon on the market, such as Mexican or Russian; they do not have the distinctive flavor of the French variety. Tarragon requires lots of warmth and sun to thrive. In container gardening, the pot must be well draining. Tarragon dislikes overly moist or rich soil. It grows irregularly and tends to drape its delicate leaves over the sides of the pot and not have the good upright habit it does when grown out in the garden. Typically, French tarragon lasts only one season in a mixed container. Transplant it to its own container with a sharp-draining soil mix.

Thyme, French (*Thymus vulgaris*): Perennial. Thyme can be easily started from seed, but I really like the impact a more mature plant gives when designing containers. Thyme is a favorite filler for containers, and this one has a look different from some culinary varieties. French thyme has long, narrow dark green leaves that create a soft, lacy mound. It can be used along the edges of pots to tumble over the sides or to fill empty pockets of space between tall herbs. French thyme (*thym* in French) is one of the stronger culinary types of thyme and one of the essential ingredients in a traditional bouquet garni, giving the mix a robust, camphor-like aroma.

OTHER CLASSIC FRENCH HERBS TO GROW

chervil (*cerfeuil*)
chives (*ciboulette*)
fennel (*fenouil*)

CLASSIC HERBES DE PROVENCE

This traditional dried herb mix seasons an array of French culinary dishes. Gently toss ¼ cup (18 g) each of dried marjoram, basil, French tarragon, and thyme, plus 1 tablespoon of rosemary (3.3 g). Store in a sealed glass jar in the spice cabinet. Use throughout the year to season sauces and soups. Sprinkle the dry mix over fresh, chopped vegetables, or combine 2 tablespoons (9 g) of the mix with ¼ cup (60 ml) of olive oil and let the flavors infuse for about half an hour. Drizzle the oil blend over chicken or potatoes just before roasting.

WILL THE REAL FENNEL PLEASE STAND UP?

The large bulb of Florence fennel can be chopped to add fresh anise flavor to stir fry, slaw, or mixed salads. You can also use the feathery leaves for fresh seasoning.

Of the two types of fennel grown in herb gardens, only the tall, leafy varieties are suitable for containers. These are commonly known as sweet fennel (*Foeniculum vulgare dulce*) and bronze fennel (*Foeniculum vulgare dulce* 'Rubrum'). Their soft, lush foliage has a light licorice aroma and flavor. Easy to start from seed, they make a tall plume accent in container gardens. I especially like to use bronze fennel to create a soft, dark cloud in the middle of a mixed planter of green, leafy herbs.

Bulb fennel, called finocchio or Florence fennel (*Foeniculum vulgare* var. *azoricum*), produces a large, fleshy white bulb at the base of the plant. Its large taproot needs so much space, this type of fennel does not typically do well in containers. It really dislikes having its roots disturbed and should be planted out in the garden in the spring after the last frost. The bulb of Florence fennel is ready to harvest when it is about the size of a tennis ball. If the bulb gets too big and mature, it may become stringy and develop an unpalatable texture.

Intense and aromatic, exotic herbs used in Thai dishes do well in containers. Cilantro, lemongrass, and variegated thyme, along with a squeeze of lime from a potted citrus tree, combine to create unique flavors. The plants are pungent and robust, but the similarity of leaf shapes and colors are best displayed in simple yet elegant pottery. A leafy setting for the pots will also boost style. Think of a sunny patio with a bamboo screen as a backdrop or a sleek wooden deck with Asian-style lighting and furniture.

PLANTS NEEDED

basil (Thai)
cilantro
lemongrass
lime (Persian)
thyme (variegated lemon)

Herbal profiles

Basil, Thai (*Ocimum basilicum*): Warm season annual tender to frost. Thai basils have a deep clovelike flavor that is distinctly different from the large-leaf Italian basils. They are popular because their pungency holds up to robust ingredients, such as garlic and curry, used in Thai cooking. Thai basils are also very decorative in container gardens. Most cultivars have burgundy stems and deep burgundy flowers. They continue to produce leaves even as they bloom, so there's no need to pinch the blooms off regularly as you do with common basils. Grow exceptionally showy varieties, such as 'Cardinal', 'Oriental Breeze', or 'Cinnamon'.

Cilantro (*Coriandrum sativum*): Annual. Grow the named cultivars 'Santo' and 'Slow Bolt' for longer container life. They produce abundant leaves and do not go to seed as quickly as common cilantro. As plants begin to bolt (the stems lengthen and produce fernlike foliage that "bolts" into flower and coriander seed production), remove them and replace them with young cilantro to keep the container fresh all season. Harvest leaves when they are young and tender. Older leaves have an unpleasant, bitter flavor.

Lime, Persian (*Citrus* x *latifolia*): Leafy, tender tree. A dwarf form of lime that tends to be a better producer of fruit. Tender in most areas, citrus outdoors does best in mild climates not prone to freezing weather. Citrus trees are a rarity and a treat when they produce fruit for most gardeners in colder climates. They are ideal for container gardeners because they can be placed strategically in a sunny, protected location in the garden and then brought indoors during the cold. Plant them alone in their pot to allow the root system plenty

of space. They produce better if you fertilize throughout the growing season. Be patient and care for a citrus plant, and it will reward you with fruit—not bushels but enough to add a fresh squeeze of lime to cooking.

Lemongrass (*Cymbopogon citratus*): Tall, grasslike tender perennial that will not tolerate frost. Typically grown as an annual in areas with long, warm summers. Specialty Asian markets usually have large viable pieces of lemongrass that can be rooted in water. Most will produce roots within a month and can be planted in soil in a pot. Three to five stalks in a small container create a lush grouping in a mixed planting; their linear structure contrasts well against leafy herbs. Use the leaves to lightly flavor food, and, as the plant base spreads, harvest the leaves down to the white base for the best flavor.

Thyme, variegated lemon (*Thymus* x *citriodorus* 'Variegata'): Hardy herbaceous perennial. Can remain partially evergreen in warmer areas. The delicate gold-edged leaves of this variegated form of thyme add texture to the sides of a container. As it grows through the season, it fills in around other plants, mingling well. Tidy and trim as needed to shape, or let the plants spill over the edges. As beautiful as it is in container gardens, it is also one of the more flavorful thymes, its strong lemon aroma touched with an earthy note.

Grow more Asian flavor

Asian foods and flavorings are diverse and encompass several cultures, including Korean, Japanese, Chinese, Vietnamese, and Thai. Their spicy flavors mix and mingle in captivating ways, and many of these cuisines rely on fresh greens to add tanginess. You can explore this in your own container gardens with interesting greens that are as beautiful as they are flavorful. The unique textures make nice annual tuck-in plants around perennial herbs. Use leafy greens around the edges of large planters or the upright stalky look of plants such as closed-head Chinese cabbages as a mid-layer. You can plant most of these in the spring to freshen up containers before other plants have come out of dormancy. After harvesting spring greens, switch them out for basil plants to fill in the gaps later in the season.

Plant starts may be hard to find in garden stores, and most are better direct-seeded into containers. See the Resources section on page 181 for a list of seed suppliers. Look for spicy-flavored greens, such as mustard green varieties 'Wasabina', 'Mizuna', or 'Scarlet Frills'. Bok choy, also known as pak choi, has large, rounded leaves that add big texture to potted gardens. Tatsoi and kale have bold, leathery leaves that can be left in containers (and beyond edibility) to add their interesting leafy texture to mixed plantings. Rocket (arugula), orach, and Malabar spinach all have tangy leaves for use in salads.

OTHER CONTAINER HERBS FOR ASIAN FLAVOR

coriander (Vietnamese)
garlic chives
mint
oregano
shiso

TULSI OR THAI?

Tulsi, or holy basil, (*Ocimum tenuiflorum*, synonym *Ocimum sanctum*) is often confused with Thai basil and sometimes labeled as such. They are two different plants with distinctly different flavors. The more common Thai basil (*Ocimum basilicum*) is used in Thai cooking for a licoricey, almost sweet, pungent flavor. The true Thai basils make beautiful ornamental plants for container gardens. Their typically abundant blooms are deep purple or smoky blue. Not as fussy or prone to disease as common basils, they will thrive in the heat of summer and tend to bloom all summer long.

Tulsi basil has a regal history, prized for use in ceremonies and temples. It is said to be popular as a plant grown around Buddhist temples in Thailand and considered a sacred healer in India. With smaller leaves and pale purple to white airy flowers that elongate above the leaves, this basil is less showy than the Thai type in container gardens. Grow it for its sharp, hot flavor—a small taste of a tulsi leaf will numb your tongue with its pungency. It is less sweet than Thai basil.

Tulsi, or sacred basil, can vary in leaf color by variety but is recognized by its deep-toothed leaf margins.

You can display rare and unusual culinary herbs, such as temperate ginger and citrus trees, to their best advantage in a container garden and move them to a protected area when frost threatens.

Ginger (*Zingiber officinale*): Culinary ginger is a tall, narrow-leaf plant with an edible rhizome that is dried, peeled, grated or sliced, and used as a warming spice in desserts and savory dishes. Culinary ginger grows easily in warmth and humidity but is not very showy in flower, so plant it in the back of containers for its leaves to provide a tropical look. It's suitable for mixed planters because it likes well-draining soil with average moisture, like most herbs.

Place the rhizome in moist soil in a pot about 1 to 2 inches (2.5 to 5 cm) deep. Look for small knobs on the rhizome. These "eyes," which are where the stems will emerge, need to be placed pointing upward just under the soil surface. For the best flavor, grow the plants for at least ten months, protecting them from frost as needed. When the leaves start to die back, pull out the whole plant. Cut off the leaves, shake off the dirt, and allow the rhizomes to dry. You can break off and replant small pieces that have "eyes."

If you prefer an ornamental ginger over a culinary one, look for showy, fragrant-blooming hardy ginger (*Hedychium* spp.). Hedychiums like moist soil and need large pots to fill with their thick rhizome system; you won't be harvesting these. Both types of ginger grow best in warm places and appreciate protection from hot sun and windy conditions. Tender to frost, gingers should be treated as an annual or brought indoors to protect from frost.

Citrus (various species): Container-grown citrus will not give you baskets of fruit, but it will give you the pleasure of intensely fragrant flowers followed by lemons, limes, or other citrus varieties. Most citrus is not hardy to freezing or prolonged cold temperatures. Containers offer mobility to protect citrus for the best chance of harvesting fruit. Keep potted plants in a sunny, warm, frost-free location outdoors and move them indoors, if needed, to protect from freezing temperatures. Plant citrus in well-draining

Citrus blooms have a heavenly scent.

pots and water regularly. Citrus needs a good, regular feeding routine to keep plants healthy in containers. Use liquid fertilizer during weekly watering in the active growing season. Dwarf varieties, perfect for containers, include 'Improved Meyer' lemon, 'Eureka' lemon, 'Bearrs' lime, and 'Nordman Seedless' or Meiwa kumquat.

Topiary citrus trees kept in pots stay transportable for trips back into the conservatory in cold periods of the year.

Saffron (*Crocus sativus*): A treasured, delicate spice, saffron is made from the dried stigmas of autumn-blooming crocus. The plant, growing from a bulb-like corm, flowers a pale shade of lilac and has dark-reddish stigmas. Because of its tiny threads, it must be hand picked and handled carefully. With about 225,000 stigmas in a pound of saffron, it is the most expensive spice in the world. You can grow your own in well-draining pots.

You will most likely find the corms available for purchase in the fall, which is when you should plant them. Place three to five bulbs about 3 inches (8 cm) deep in the soil in a 12-inch (30 cm) terra cotta pot. They typically do not bloom the first year they are planted but will the following fall season. You can also tuck bulbs into mixed herb containers, giving them enough space to emerge around plants in the fall. They make a good addition in the space where you've harvested out annuals, such as basil. Once they are in flower, harvest the stigmas carefully and dry them on screens.

MICROGREENS! TINY PLANTS, BIG HERB FLAVOR

Still waiting for your parsley to leaf out or the basil to get big enough to eat? Extend your season of flavors with a micro herb garden. Microgreens are simple; they are herbs grown from seed and harvested when they are about 2 inches (5 cm) tall. These are tiny plants packed with flavor, and I like to have trays of microgreens on hand, especially in the off season, when I can't go outside and harvest fresh from the garden.

You can grow many herbs, flowers, and vegetables as microgreens for flavorful, bite-sized treats. Use them to power up a salad, top warm vegetables just before serving, sauté in stir fry, and replace lettuce on a sandwich. The little bursts of fresh flavor do not overwhelm the taste of the main dish.

Growing microgreens indoors is a quick way to have the flavor of herbs without having large pots of single herbs on the windowsill. With the right setup, you can grow them year round, and most are ready to harvest in two to three weeks. Sowing small trays every few weeks will give you a steady crop of flavor.

Herbs for microgreens

The best herbs for microgreens are ones that germinate reliably and quickly and have a lot of flavor in their leaves.

Basil (*Ocimum* spp.): All basil cultivars work as microgreens. Try these for best flavor: 'Genovese', lemon, and Italian large-leaf. For miniature pops of standout color, use purple-leaf varieties: 'Red Rubin', 'Dark Opal', and 'Amethyst'.

Borage (*Borago officinalis*): Quick germinating with big, thick leaves that have a cucumber flavor

Cilantro (*Coriandrum sativum*): Lacy, delicate green leaves with a strong cilantro aroma and flavor

Dill (*Anethum graveolens*): Light, feathery leaves with a mild, fresh dill flavor

Microgreen herbs are easy to grow and add tons of flavor to meals.

Fennel (*Foeniculum vulgare*): Delicate, narrow leaves with a subtle licorice flavor

Lemon balm (*Melissa officinalis*): Slower growing than most microgreens, lemon balm as a microgreen will add a lemongrass-like flavor and aroma to food.

Parsley (*Petroselinum crispum*): Slow to start but one of my favorite microgreens; the flavor of parsley is more intense as a microgreen.

Shiso (*Perilla frutescens*): Tiny, mildly spicy clove flavor; grow red-leaf shiso to add color to a microgreen mix.

Diversify your flavor mix by adding vegetable greens to your microgreen garden. Lettuce, kale, mustard, and other greens are all fast-growing plants you can easily grow as microgreens. Radish, mustard, cress, and arugula have a spicy bite. 'Lemon Gem' marigold is typically grown as an edible flower but makes a good microgreen: The leaves have a citrusy lemon peel-like taste and aroma.

How to Grow Microgreens

Fill a shallow tray with 2 inches (5 cm) of soil mix. Sprinkle seed evenly over the tray. Try to cover the soil surface as much as possible; fill all the space. It doesn't have to be perfect, you just don't want the seeds clumped together on the soil surface. Lightly press the surface to ensure the seeds have made contact with the soil. Small seeds do not need to be covered with soil. Large seeds the size of sunflowers or peas need to be covered with a thin layer of soil.

Water gently with a misting spray bottle. To avoid disturbing the seeds, do not pour water directly over the tray. Keep the soil consistently moist and never let the tray dry out.

SUPPLIES

Shallow trays: Plant flats used for starting seeds work well; the trays need to be 2 to 3 inches (5 to 8 cm) deep. Look for small trays that fit in the space you have.

Soilless or seed-starting soil mix: Most seed-starting mixes are light and hold moisture well. Make sure the mix is thoroughly moist before adding it to the planting tray. You want a fluffy, damp mix in which the seeds can easily germinate.

Seeds: Herb seeds that you use to grow plants for the garden are the same ones to use to grow microgreens. Just make sure to look on the seed packet label to make sure they are untreated and organic.

Misting spray bottle: Most of these seeds are very small. A misting spray will keep them watered without washing them away.

Seeds germinate best when the soil is kept consistently warm—about about 65°F to 70°F (18°C to 21°C). Seedling heat mats are handy to help with this, or place your trays on top of a home appliance, such as a clothes dryer, that generates heat. Once the seeds germinate, your baby plants will need proper lighting to grow healthy, flavorful leaves. A south-facing window with good light is best. Use a small lamp with a grow light bulb to add lighting if you need it.

Harvest microgreens when the cotyledons (the first leaves that appear from the seed) have fully developed and you begin to see the first true leaves emerge. Most microgreens will be about 2 inches (5 cm) tall. Cut the greens off at the soil level and give them a light rinse before eating.

CHAPTER 5

HERBAL BEVERAGES

FLAVORFUL AND SWEET HERBS
FOR TEA MIXES AND COCKTAILS

How to drink your herbs? It's all about infusion! Steeping plants into a liquid will infuse the essence of the plants into a pourable, drinkable form. That liquid captures all the attributes of the herbs, including their flavor and aromatic notes. Make flavorful healing teas; infuse fresh fruit and herbs into water for delicious, healthy hydration; and muddle the pungency of herbs like basil and mint into unique cocktails.

The two herb container projects in this chapter offer herbal selections for tea blending and fun edibles to grow alongside your herbs for creating inspiring cocktails.

Tea has a very long and storied history. We think of tea both as a hot, nurturing beverage and as a refreshing, iced drink in the summer. True tea is made from the leaves of *Camellia sinensis*, a large evergreen shrub grown in temperate climates, such as China and India. Tea growers harvest the leaves and process them in myriad ways, such as fermenting, drying, and curing, to create green, oolong, black, and other familiar types of tea.

The word *tea* has come to refer not only to *Camellia sinensis* but also more broadly to other herbs used to infuse hot water with the essence of the plant's aroma and flavor. Grow both the actual tea plant, *Camellia sinensis*, and other richly aromatic herbs to create delicious blends from the garden that capture all the healing, flavorful qualities of the plants.

A tea garden is the perfect setting for quiet contemplation. Add containers to an area with comfortable seating surrounded by calming colors. Harvest small amounts of leaves and flowers throughout the year to use fresh. At the end of the season, when plants are going into winter dormancy, harvest leaves and dry them for use throughout the year.

HARVEST AND DRY HERBS FOR TEA

Herbs are at their most flavorful and fragrant in the morning after the dew dries but before the hot sun releases their essential oils. For tea blending, it is best to dry the leaves and flowers whole and then break them down to release their preserved flavors. Dry small leaves and flowers on drying screens.

Herbal profiles

Chamomile (*Chamaemelum nobile*): Herbaceous, spreading hardy perennial. Roman chamomile is a fragrant perennial carpet. Use this low, mat-forming plant in container gardens as a living topdressing to fringe the edges of pottery. Use scissors to shear lightly as needed. The small daisy-like flowers appear throughout the growing season and are the main plant part used for tea.

Lavender 'Hidcote' (*Lavandula angustifolia* 'Hidcote'): Herbaceous perennial. There are hundreds of varieties of lavender to grow, but the sweetest flavor is from the dark purple varieties of English lavender. 'Hidcote' is a compact form with abundant deep purple blooms in the midsummer. Harvest the flower buds and use fresh or dried in teas. The earthy flavor of lavender combines well with lemon verbena and chamomile for a calming, aromatic tea blend.

Lemon verbena (*Aloysia citriodora*): Tender woody perennial. Lemon verbena is sensitive to frost and is typically treated as an annual in cold winter areas. It is a woody, shrub-like plant that can grow up to 8 feet (2.4 m) tall if left untrimmed. Keep it nice and tidy in a container by trimming leaves and flowers to encourage full, bushy leaf growth. Harvest the leaves during the summer; they are aromatic when freshly picked and one of the best lemony herbs you can grow. Harvest all the leaves from the stems if frost threatens, to preserve for use in the winter.

Sage, pineapple (*Salvia elegans* 'Golden Delicious'): Herbaceous perennial that is tender to frost and typically grown as an annual. Tall and dramatic in container gardens, pineapple sage has a color and growth habit that makes it a good, showy contrast plant to add height in containers. Harvest leaves for a fragrant, fruity addition to tea mixes.

Spearmint (*Mentha spicata*): Hardy herbaceous perennial. Spearmint is a favorite for teas because it has a sweeter flavor than common peppermint. Mint can be kept in a mixed container with other plants for no more than a season. Due to the aggressive nature of its root system, spearmint will soon overtake other plants. Remove and divide it, then plant one division back into the original space for another growing season. Trim plants regularly and remove the flowering parts to keep leaves going through the season. New leaves have a more robust flavor than older ones.

Stevia (*Stevia rebaudiana*): Annual. Best to try and find a plant at the nursery, but can be started from seed. Stevia will grow rangy and needs trimming a few times over the growing season to keep an attractive shape. Remove flowers to promote leaf growth and keep it tidy. If you are in an area prone to slugs, keep a watch out, as they love the sweetness of the leaves. Known as the "sweet leaf" herb, stevia contains stevioside, which is said to be two to three hundred times sweeter than sucrose. Harvest leaves for use any time, but they will be at their sweetest just before the plant flowers. Crumble the fresh or dried leaves in tea blends to add natural sweetness.

OTHER TEA HERBS FOR CONTAINER GARDENS

lemon balm
rose-scented geranium
rosehips
roses
tulsi basil

Sweet woodruff (*Galium odoratum*): Hardy herbaceous perennial. A sweetly scented, low-growing plant. It prefers a little bit of protection from the sun and is perfect as topdressing under more abundant herbs that will shade it. Star-like white flowers appear in the spring, and the leaves have a sweet, vanilla-like fragrance. Steep the leaves, fresh or dried, in hot water for a slightly sweet, aromatic drink that promotes relaxation.

Tea plant (*Camellia sinensis*): Evergreen shrub. *Camellia sinensis* is an attractive glossy evergreen that makes a beautiful tall statement in a container garden. It is one of the Chinese varieties that is hardier and can take light frosts but needs protection in areas prone to freezing winters. Camellia is best by itself in a large pot (shown here with a low ground cover for interest as an understory around the edge of the pot), so it can grow truer to its natural, upright shrubby habit. This camellia grows up to 6 feet (1.8 m) or taller but is easily maintained at 3 to 4 feet (0.9 to 1.2 m). Regular trimming and harvesting will help keep the plant an attractive shrubby form. Tiny white flowers appear up and down the branches in the fall. Harvest the young tender leaves in the spring for fresh green tea.

A FRESH POT OF TEA

Boil water in a teakettle. Transfer boiling water into a warmed ceramic teapot. Add fresh or dried herbs—1 tablespoon fresh or 1 teaspoon dried to 1 cup (235 ml) of water—and allow to steep for at least ten minutes, longer for a stronger tea. Judge your herb tea by taste rather than by color, because most become a light greenish amber rather than a deeper color.

Plant your recipe ingredients for a summer cocktail party. Beverages, whether a cocktail mixed with alcohol or a mocktail without liquor, go to a new flavor zone when infused with herbs. This fun container mix gives you the pungency of herbs with the sweetness of edible berries. The berry cultivars with trade names (Jelly Bean®, Raspberry Shortcake®) are bred to stay compact yet highly productive in container gardens to give you plenty of fruit to blend in drinks and use for garnish.

Pop in colorful plant additions, such as annual flowering herbs, violas, borage, and nasturtiums. Venture into growing citrus, such as Meyer lemon, in containers. Tuck these pretty containers into a garden border along the walkway for easy access. A carpet of variegated pineapple mint (*Mentha suaveolens* 'Variegata') planted in the ground underneath makes the pots look naturally nestled into the space.

PLANTS NEEDED

basil
blueberry, Jelly Bean®
lavender 'Hidcote'
raspberry, Raspberry
    Shortcake®
scented geranium 'Lady
    Plymouth'
spearmint

The bright blue edible flowers of borage have a cucumber-like flavor. Plant borage for annual color in a container and harvest the fresh flowers to float in your herbal cocktails.

PLANTS FOR GARNISH

Decorating and embellishing food and drink is like adding jewelry to complete an outfit. Adding fresh herbs as a garnish somehow makes great-tasting dishes (especially those that may look a bit boring) even more appetizing. Many herbal flowers and leaves make beautiful garnishes, particularly for cocktail glass rims and to suspend in ice cubes. Grow these herbs for their edible and beautiful leaves and stems: spearmint, variegated mint, large-leaf sweet basil, flat-leaf and curly parsley, rosemary (tender stems), lavender (stems with buds), sage, and savory. Borage, chive, chamomile, nasturtium, pansy, and viola all have edible flowers.

"At last we venture in the garden, where we encounter a seasonal array of botanical mixers and garnishes."
—AMY STEWART, *THE DRUNKEN BOTANIST*

Herbal profiles

Basil (*Ocimum basilicum*): Annual. Grow sweet basil and lettuce-leaf basil for sweetness combined with a classic basil flavor. Purple-leaf basils, such as 'Dark Opal' and 'Red Rubin', add a pink color to drinks. Mash the leaves to flavor gin and vodka for a refreshing kick of flavor. Use the leaves to float as a garnish. Smash the leaves into a nonalcoholic Lime Rickey (lime juice, simple syrup, and seltzer) for a refresher on a hot summer day.

Blueberry, Jelly Bean (*Vaccinium corymbosum* 'ZF06-179'): Deciduous shrub. A compact selection with beautiful blue-green leaves tinged with red in colder weather. The plants have showy, pinkish-white flowers in the spring followed by sweet little blueberries about midsummer. This blueberry is part of a newer line of plants being bred for container gardens and small spaces. It stays under 2 feet (0.6 m) high and wide and makes a good filler in the middle or back of a larger planter. Use alone in a medium-sized pot for a nice anchor to a trio of containers.

▼ **Lavender 'Hidcote'** (*Lavandula angustifolia* 'Hidcote'): Herbaceous perennial. Compact English lavender with deep purple flowers. This one is a favorite in culinary dishes because the flower buds have a sweeter flavor than other lavenders. Its tidy growing habit makes it a good companion mixed into planters. For best flavor, harvest the buds when they turn from light gray to deep purple and before the flower petals open up. You can use them fresh or dried, and fresh-cut stems with the flowers left on make a fun swizzle stick.

Raspberry, Raspberry Shortcake (*Rubus idaeus* 'NR7'): Deciduous shrub. Compact and rounded, this is part of the plant breeding program bringing container garden edibles into the home gardening market. If you are familiar with growing raspberries, you know they take a lot of space and some type of trellis in a garden. These cultivars grow rounded and compact without the fuss of staking, just right for a balcony or deck planter. Thornless stems make them pleasant in a pot around living spaces; they won't snag clothing or skin as you walk by. The only challenge I have with this plant is that birds love the berries and sometimes get them before I do.

OTHER HERBS FOR A COCKTAIL CONTAINER GARDEN

lemon verbena
rosemary
sage
stevia
thyme

Scented geranium 'Lady Plymouth' (*Pelargonium graveolens variegata*): Tender perennial. Treat as an annual or bring inside in areas prone to freezing temperatures. Lovely leaf texture makes this a favorite scented geranium variety for use in container gardening. The flowers are small and light pink, but it almost doesn't matter because the leaves are the showy part of the plant. It stays well behaved in containers; snip off spent flowers and stems to keep a tidy appearance. If the plants get too shaded or crowded by other plants, they may get leggy. Trim the elongated green stems to the next leaf growth area, and they should flush out nice and full again. Use the fresh leaves as a garnish in a drink. They add an earthy aroma when muddled into liquids, or you can infuse the leaves into a simple syrup to use as a cocktail mixer.

Spearmint (*Mentha spicata*): Perennial. Like many plants in the mint family, spearmint has an aggressive root system that is best planted in its own pot so that it won't crowd out other plants. Light green, spear-shaped leaves have a sweeter flavor than common peppermint. The sweetness combined with its robust growth habit makes this a plant you can trim regularly and still have enough for teas, cocktails, and more. This is the classic herb leaf muddled into Kentucky bourbon for mint juleps.

SIMPLE SYRUP

Simple syrups sweeten and add flavor to cocktails and teas. Unlike granular sugar, which takes time to dissolve, simple syrups quickly blend into drinks. You can use almost any herb, but infuse them individually for the best flavor.

> 1 cup (235 ml) water
>
> 1 cup (200 g) sugar
>
> 5 large fresh leaves basil (can substitute ¼ cup [18 g] fresh lavender buds OR [5] 5-inch [13 cm] tender stems spearmint)

Bring sugar and water to a boil over low heat in a glass or non-reactive saucepan. Stir until sugar is completely dissolved. Remove the sugar water mix from the heat. Stir in the herb of choice and allow the syrup to completely cool. Pour the syrup through a mesh strainer and into a sterilized glass jar or bottle. Store in the refrigerator and use within two weeks.

CHAPTER 6

HEALING HERBS

COMPACT HERBS TO GROW FOR
HOME REMEDIES AND AROMATHERAPY

I am fascinated by the history of herbs and their uses. I like reading about the origins of their names and how humans use them to care for each other. It seems that whatever the ailment, there was a plant to cure it. Long before we had a chance to walk the aisles of a drugstore, the garden was the apothecary—a poultice of this, a tincture of that ... something for everyone. And since antiquity, aromatic plant essences have been understood to have beneficial effects on mental and emotional health. You can grow herbs in containers for medicinal and aromatherapeutic purposes, bringing a healing garden within easy reach.

MEDICINAL HERBS

The long-standing tradition of using herbs for healing is an easy study in your own backyard. It begins with an exploration of how particular herbs heal and what part of the plant to use. Medicinal herbs are not always showy, but there are some that add decorative magic to container gardens, and you can choose interesting containers for them. Elegant metal urns show off the healing herbs in the design opposite, all of which have been used to calm and soothe. You can always add seasonal color with calendula or burgundy-leaf basils, such as 'Amethyst Improved'. Find a relaxing place in the garden to set your potted apothecary, perhaps near a seating area or an outdoor bathing space. Harvest fresh herbs as needed for a quick healing poultice or a soothing cup of tea.

PLANTS NEEDED

ginkgo 'Mariken'
lavender 'Hidcote'
lemon balm
rosemary 'Irene'

Herbal profiles

Ginkgo 'Mariken' (*Ginkgo biloba* 'Mariken'): Deciduous tree. The small-growing variety 'Mariken' is a slow-growing, semi-weeping form. Easily trained as a small topiary, it provides a taller sculptural element to container gardens. The distinctive leaf shape adds unique interest in the spring and summer. In the fall, the leaves turn brilliant yellow before they drop; the silhouette of the leafless branches and trunk becomes an interesting sculpture in the winter months. Gingko is one of the herbal trees treasured for its long history as a medicine. In European and traditional Chinese medicine, gingko leaf and its extracts are among the most popular herbs on the market. Respected studies have shown that ginkgo leaves used for medical purposes improve brain efficiency and memory. In your home remedy garden, the leaves can be dried and used to make a calming bedtime tea.

Lavender 'Hidcote' (*Lavandula angustifolia* 'Hidcote'): Perennial. 'Hidcote' is a compact variety of classic English lavender. The silvery-foliaged plants fill in gaps between other leafy herbs in mixed potted gardens. The entire plant is highly fragrant with essential oil, but the greatest concentration of healing properties is in the flowers. Cut the stems when the flower buds are not yet open but have a deep purple color. Tidy the plant to a well-rounded shape after harvesting the flower stems. Sometimes the plant will put on a second bloom late in the summer and fall. Bundle the harvested stems and flowers and hang to dry. Use the flower buds in tea, bath salts, and sachets. English lavenders (*L. angustifolia*) are traditionally revered over other species for use as medicine. They are super healers with a reputation for treating burned skin, relieving pain, stopping migraines, and so much more. A sprig of fresh lavender tucked behind your ear or under a hat on a warm day will soothe a headache and relax you after a stressful day.

Lemon balm (*Melissa officinalis*): Perennial. Spreading, with a height up to 3 feet (0.9 m). Best planted by itself as the plant's root system will rapidly spread and fill a container. All parts of lemon balm have a strong fragrance of soapy lemons. The leaves, both fresh and dried, are popular for use in healing remedies. A simple infusion in hot water becomes a medicinal tea to help alleviate a headache, calm nerves, and promote restful sleep. The healing water also soothes mouth sores and can be used as a sore throat gargle. In studies, lemon balm's essential oil components have been shown to help with depression and nervous anxiety. In medieval times, lemon balm was a popular strewing herb: The plants were folded in linens and in herb pillows for a cheering effect in the household. Harvest leaves from your container garden just as the plants go into dormancy, and dry the leaves. Use them in potpourri for a refreshing fragrance that cleans stale air in the winter.

Rosemary 'Irene' (*Rosmarinus officinalis* 'Irene', synonym *Rosmarinus officinalis* 'Renzels'): Perennial. Rosemary loves hot, dry summers and will flourish in a well-draining pot in full sun. Keep a close watch on how you water potted rosemary. It despises being too wet and its roots will easily rot, but if the soil dries out, it will drop its leaves and die. Lovely and aromatic, creeping varieties, such as 'Irene' and 'Huntington Carpet',

can be used as spillers over the sides of containers and in hanging baskets. The delicate blue flowers bloom off and on through the winter in milder areas. Note: Creeping varieties of rosemary tend to be less hardy than upright cultivars and may die in freezing temperatures. Treat as an annual in regions with long, very cold winters. The narrow, resinous leaves are the part of the plant used in healing remedies. A tea sipped after a heavy meal will aid digestion. Rosemary is also known for its pain-relieving properties. Make a strong infusion of fresh leaves in sweet almond oil and use it to massage away joint pain.

Calendula flower petals drying for use in blending as a healing salve.

OTHER HEALING HERBS FOR CONTAINER GARDENS

aloe vera (healing burns, skin)
basil (calming highly irritated skin)
calendula (calming, soothing skin)
chamomile (calming tea promotes sleep)
mint (stimulating, uplifting)
sage (healing, anti-inflammatory)
thyme (healing, antiseptic)

Tips for using healing herbs as home remedies

You can further explore using herbs from your garden to heal by becoming familiar with the centuries of the history and study of their medicinal uses. Start with a few plants, such as the ones I have suggested for use in containers. Research everything you can about each plant, what it is traditionally used for, and its safety precautions. Use reputable sources to discover how and what the herb heals, which parts to use, and in what formulation. Check the Resources section on page 181 for some of my favorite, trusted references. Some herbs are not suited for internal use, and some are okay in less concentrated formulations. Choose wisely, know your own health, and respect the power of plants in your garden.

Once you discover the healing attributes of garden herbs, one of the easiest ways to begin using them is by infusing them into hot water. Infusion might be more familiar to you as the process of making tea. Place fresh or dried leaves in hot water to release all of the plant's qualities into a liquid, useable form. Depending on the herb and its intended use, the water can then be sipped as a tea, used as a skin tonic, or mixed with other ingredients to create a poultice or lotion.

MIDSUMMER HEALING TEA

In midsummer, when lavender is ready for harvest, cut the stems with flower buds that are deep, dark purple, just before the petals open. Hang the bundled stems to dry. Cut a handful of lemon balm (about ten leaves), a ginkgo leaf, and a few stems of rosemary and place them on a screen for drying. Once all the plants are dry, strip the lavender buds and rosemary off the stems. Crush them together with the dried lemon balm and ginkgo until all the larger pieces have broken down. Toss gently to mix well, then infuse into hot water. The healing properties of the herbs combined in this tea mix will help you relax after a busy day.

Topiary herbs

The container garden shown below holds an unusual form of dwarf ginkgo that is trained into topiary. This plant was a chance find at a local nursery and a must-have for my healing remedy container garden. *Topiary* is a term that describes an elegant form of training plants that is also a good way to get height in container gardens. It is basically a way to make a plant look like a miniature tree. The lower branches are completely removed from a straight stem to form the "trunk." The leafy part of the plant is typically left as a round or loose ball shape at the top. The top is shaped regularly to keep its original form. Most herb topiaries in containers are easy to keep. A simple haircut once or twice in the growing season usually maintains their shape.

Many woody, shrublike herbs can be trained or purchased as topiary. Rosemary and Spanish lavender are two of the herbs most used as topiary because their woody stems make for easy trimming, they fill in quickly, and they hold their shape with regular maintenance. Other herbs typically clipped into topiary include germander, scented geraniums, myrtle, and Australian bush mint (*Prostanthera*). Check local garden stores and herb shops for topiaries; rosemary ones tend to show up during the holidays for seasonal decorating.

IS BOXWOOD AN HERB?

As plants are defined for their usefulness and called "herbs," there are many obscure healing claims attributed to ornamental plants. Boxwood, a common plant used in topiary design, has an interesting reputation, perhaps, long forgotten or possibly debunked in our modern-day usage. Seventeenth-century herbalists made great claims for the medicinal use of *Buxus*. As noted in *A Modern Herbal* by Mrs. M. Grieve, a decoction of the leaves when applied to a bald head would promote the growth of hair. Other herbalists suggested using the oil in the leaves, woody stems, and roots as a cure for leprosy, a hair dye, and a treatment for epilepsy; the wood is said to have narcotic properties.

Boxwood trimmed as a topiary adds interesting container garden texture to a potager garden.

*A*romatherapy is an obvious combination of the words *aroma* and *therapy*. A more in-depth consideration of those two words yields an understanding of what scent can do for well-being. Fragrance is a powerful sensory experience. The scent of cinnamon, for example, can transport someone to their childhood or a holiday celebration, while the perfume of a flower can call to mind a special person or place. The human nose can recognize more than ten thousand scents and more than three thousand chemical components that create different fragrances. Numerous plant families create such scents, but the majority are members of the vast family called herbs. It is one of the most recognizable characteristics that enchant us into growing and using them.

In a garden, fragrant plants play a significant role. Crushing chamomile releases the smell of green apples, and lemon thyme rubbed between your palms produces an uplifting citrus note. Rosemary and hyssop releasing essential oils in the hot summer sun not only draw bees and other insects, they get our attention as well. Walk around a lavender farm on a sunny day and the aromatic oils in the lavender take hold of your senses and naturally relax you.

Container gardens are an opportunity to create concentrated groupings of plants that make intense perfume. Placement is vital; you want the plants where you can brush up against them or have the aroma float around you on a hot summer day. The intention of container gardens filled with aromatic herbs is to position fragrance where you can really catch it: in window boxes so the scent drifts in through open windows, along pathways and outdoor living spaces to add an aromatic dimension to the garden. Walk through your garden and imagine the best spot for the fragrance to surround you.

You can also make a traditional dry potpourri from the herbs in this container. Create sachets to impart fragrance in linen closets and dresser drawers. Tuck them into suitcases while in storage to keep them fresh for their next use.

> **PLANTS NEEDED**
> lavender (Spanish)
> roses
> scented geranium 'Old Fashioned Rose'
> scented geranium (peppermint)
> thyme 'Foxley'
> thyme (variegated lemon)

POTPOURRI

Harvest geranium leaves, rose petals, and lavender buds and dry them on screens. Once they are completely dry, gently mix them in a gallon glass jar. If desired, you can add up to 10 drops of lavender essential oil and gently shake to blend well. Seal the jar and place it in a dark cabinet for a few weeks so the essential oils of the herbs create a fragrant blend. Place in fabric bags.

Herbal profiles

Lavender, Spanish (*Lavandula stoechas*): Perennial. Spanish lavender is also commonly called French or butterfly lavender. It is noted for its fat, tufted flowers, which look like bumblebees hovering over the plants. Its small, pointed silver leaves are highly aromatic. This popular Mediterranean herb grows best in full sun. In areas of the world where it is native, it thrives in rocky, lean soils. Container-grown lavender needs to be very well drained. The plants will appreciate soil that is not overly rich with compost or fertilizer. Keeping soil lean and well draining will make the essence of lavender stronger. The stronger the oil, the better the aromatherapy. Lavender used in a garden space will lend its calming scent to the air as the hot sun triggers the release of the essential oils.

Rose (*Rosa* spp.): Deciduous shrub. Roses grow best alone in deep pots (18 to 24 inches [46 to 61 cm]) to allow roots to spread out. Keep soil nourished by top-dressing the rose container with compost every season. Good air circulation around the pot helps reduce the occurrence of fungal diseases. Add rose pots to the back of a grouping of other potted herbs to create a mass of fragrance. Grow compact, aromatic varieties that stay nice in containers, such as 'Munstead Wood', 'Harlow Carr', or true miniatures, such as 'SAVaseat' (trade name Seattle Scentsation) or 'SAVamor' (trade name Scent-sational). Growing roses in containers gives you the luxury of capturing elusive rose fragrances in small spaces. The aromatic oils of rose are relaxing and stress relieving. Gather rose petals from your containers to make fragrant natural potpourri or make rose water to use as a mist to cool down on a hot day.

Scented geranium 'Old Fashioned Rose' (*Pelargonium graveolens* 'Old Fashioned Rose') and **Scented geranium 'Peppermint'** (*Pelargonium tomentosum*): Annual or tender perennial. 'Old Fashioned Rose' has deeply cut, frilly leaves, and peppermint-scented geranium is noted for its large, soft, fuzzy leaves. Scented-leaf geraniums are highly aromatic and easy to grow in a warm and sunny location. They are tender to frost and will not survive freezing winter temperatures. Pelargoniums grow fast in the heat of summer, almost to the point of becoming gangly and unkempt. The plants typically need to be trimmed a few times through the season to keep them within bounds in mixed container gardens. Healing and aromatic, scented geraniums are known in aromatherapy as uplifting. Are you feeling an afternoon slump or a bit of brain fog? Pick the fragrant leaves to rub between your palms, then deeply inhale the aroma for a natural pick-me-up. Dried leaves also hold their fragrance, making them a good addition to air-freshening potpourris.

Thyme 'Foxley' (*Thymus pulegioides* 'Foxley'): Perennial. Thyme in a window box or planter is the perfect filler to tuck around the base of taller plants. The lazy, mounding habit of thyme will spill over the sides of containers. The variety 'Foxley' has a larger leaf than most common thyme cultivars. The color of the leaves is unique and vibrant, with creamy variegation to add an extra shot of contrast in mixed planters between green, leafy herbs. When the deep lilac-colored flowers bloom in the summer, the plant readily releases fragrance in the heat of the day. Other times of the year, a quick brush of the leaves will release the essential oils on your hand and into the air.

Thyme, variegated lemon (*Thymus* x *citriodorus* 'Variegata'): Perennial. Tiny, golden variegated leaves tumble easily over the side of container gardens and window boxes. Use thyme in living walls as an excellent filler and for its fragrance. The fresh lemony aroma and the added color on its leaves make this variety one of the most popular thymes. Lemon thyme can be dried for use in natural air-freshening potpourris. In aromatherapy practice, thyme has a refreshing, warming herbal aroma that helps "clear the air" and is noted in historical herb references for its use as incense.

PATCHOULI: YOU MIGHT KNOW THE FRAGRANCE BUT NOT THE PLANT

Patchouli (*Pogostemon cablin*) is the plant responsible for that distinctive earthy aroma used in soap and perfumes. The leaves are highly fragrant and can be dried for use in potpourri and sachets. You can grow this interesting, unusual herb in aromatherapy container gardens. It is not showy in flower but can be used as a green, leafy filler. Treat it as an annual; it grows best in warm weather and will not survive a frost. Look for plants from specialty herb growers and propagate by cuttings in the fall.

OTHER AROMATHERAPY HERBS FOR CONTAINER GARDENS

basil, Thai (invigorating)
chamomile (calming, soothing)
patchouli (balancing, relaxing)
peppermint (uplifting, energizing)
rosemary (energizing)

Lemony scents give us a lift. This tabletop garden in a small galvanized tub imparts the irresistible fresh fragrance of lemons. The multipurpose herbs in this mix include lemongrass for cooking; lemon verbena, popular in soap making; and thyme and scented geranium, which have strong scents that linger in the warmth of summer. Place a garden of these fresh lemon herbs near a seating area or somewhere you can brush your hands across the leaves to release their fragrances. An added bonus to the mix of plants is that the lemony scents are naturally deodorizing to the air. This is an easy centerpiece for a patio table.

This shallow metal tub container will be useful for only a season or two before the plants outgrow their space. You can treat this as an annual container and refresh it every season with smaller herbs for continued use as a tabletop planter. Add bright flowering annual herbs, such as 'Lemon Twist' calendula and 'Lemon Gem' marigold, to pop yellow color throughout the summer. Then bring the container into the house before the threat of frost and use it as an indoor arrangement.

> **PLANTS NEEDED**
>
> lemongrass
> lemon verbena
> scented geranium 'Lemon Fizz'
> thyme (variegated lemon)

According to research at Ohio State University, smelling lemon raises levels of norepinephrine, a brain chemical linked to easier decision-making and heightened motivation.

Herbal profiles

Lemongrass (*Cymbopogon citratus*): Tender perennial, treated as an annual in most areas. Do not grow from seed; instead look for plants at garden centers for instant height in a container garden. Not showy with flowers or colored leaves, lemongrass is best used in container design for the texture of its tall, grassy foliage. This is a true grass that has an earthy lemon aroma when brushed against or harvested. The plants have a shallow root system and will not compete with other plants for space. The secret to keeping lemongrass healthy is to grow it in a sunny spot and regularly water the container. It is native to warm, humid climates and appreciates moisture. At the end of the season or when frost threatens, harvest the leaves and braid them. Allow them to dry completely and store them in a glass jar. Snip the ends of the dried plant to flavor tea.

Lemon verbena (*Aloysia citrodora*): Half-hardy woody shrub. Tender to frost and typically grown as an annual. In a small container, the plant can be regularly trimmed to stay tidy and compact. Lemon verbena can be brought indoors and placed in a sunny, warm spot to overwinter in cold climates. When brought indoors after a summer outside, it can have a hard time adjusting and may fuss, drop all of its leaves, and go into dormancy. Don't give up; as long as the soil in the pot does not completely dry up, new leaves will emerge along the woody stems and flush out to a vibrant plant again.

Scented geranium 'Lemon Fizz' (*Pelargonium* 'Lemon Fizz'): Tender perennial. This compact, super-intense lemon-scented geranium cultivar does not grow as fast as other pelargoniums, so it is well behaved in shallow mixed containers. The stiff upright habit makes it an excellent mid-layer plant between lemongrass and lemon verbena. Light pink flowers with a deep burgundy splotch of color come and go through the warm growing season. Deadhead brown flower stems by brushing your hand or lightly shaking the plant. The remnants will fall off, and you will get a good dose of aromatherapy! The fragrance will remain in the leaves after they are dried and are a nice addition to potpourri mixes.

Thyme, variegated lemon (*Thymus* x *citriodorus* 'Variegata'): Perennial. Hardy and fuss-free. Thyme doesn't require a lot of water or fertile soil, and it is the perfect low-growing spiller in containers. The lazy, mounding habit creeps around the base of all the other plants to fill in the gaps. Because of its relatively shallow root system, it plays well in mixed containers, window boxes, and living walls. This variegated-leaf variety adds more to the lemony yellow theme and brightens up against the plain leaves of other herbs in the mix. Multifaceted thyme offers benefits in healing, skin care, and cleaning products, but it is most familiar for lending its pungency to cooking. Cut tender stems and leaves to season pork, seafood, or baked chicken.

Lemony insect-repellent plants

There has been a lot of buzz about planting lemony and citrus-fragranced plants near patios and outdoor spaces to keep away mosquitos and other pesky insects. The plant marketplace now offers one called "the mosquito plant" (*Pelargonium* 'Citrosum'). The label claims that, if it is planted in a garden, it will keep away mosquitos. This particular plant is a strongly aromatic scented-leaf geranium, not to be confused with actual citronella-producing plants, which are lemon grass species that produce the oil used in citronella candles.

Does growing plants to repel insects really work? Well, yes and no. Studies show that having these plants around helps to lessen the visiting population, but it takes a highly concentrated release of aroma to work. The plants would have to be continually crushed to be truly effective. Protection is more likely when you infuse the leaves of such plants into water and use it as a spray mist. A study shows that a mix of plants containing the chemical compounds citronella (lemongrass), linalool (lemon verbena and thyme, to name just a couple), and geraniol (scented-leaf geranium) in their leaves will have the greatest effect. So just planting them does not work, but a highly aromatic mix will give you ingredients that have insect-repelling properties.

Lemongrass and scented geranium 'Citrosum'.

A GARDEN OF LEMON DELIGHT

Ramp up your love of lemons with lemony fresh aromatic plants. Looking beyond the category of herbs, these trees and shrubs will lend fresh citrusy fragrance to the garden.

geranium 'Biokovo' (*Geranium* x *cantabrigiense* 'Biokovo'), fragrant leaves and flowers
Japanese mock orange (*Pittosporum tobira*), fragrant flowers in the summer
Mexican orange (*Choisya ternata*), fragrant leaves and flowers
mock orange (*Philadelphus* spp.), fragrant flowers, especially the cultivar 'Avalanche'
southern magnolia (*Magnolia grandiflora*), a large tree with creamy white, fragrant flowers
winter daphne (*Daphne odora*), intensely fragrant flowers in late winter

CHAPTER 7

NATURAL BEAUTY AND HOUSEKEEPING

HERBS FOR NATURAL HOMEMADE SKIN CARE AND HOME CARE PRODUCTS

I have found that one of the easiest ways to explore the use of herbs is to make natural skin and home care products. You can use the healing properties of plants for topical skin care. Heal a burn, soften dry hands, soothe irritated skin, and get rid of puffy bags under your eyes—all by simply harvesting and using herbs you can grow in your garden. Herbs from the garden can also replace an array of household cleaning products available in the store to scrub, polish, and scour. It is mind-boggling what chemicals are introduced into a daily routine by these products. If you or someone living in your home has sensitivities or you simply want to have safer and natural cleaning and skin care products, turn to herbs in containers.

The commercial cosmetic industry has spoiled us with the ease of going to the store and picking out a lotion from the shelf. We know it will smell good, have a lovely smooth consistency, and stay stable in the bottle (not separate or turn rancid). All those ingredients on the bottle that are hard to pronounce or are unrecognizable are the things that allow products to be shipped all over the world in hot or cold weather and still be store ready. People with sensitive skin are more apt to be allergic to some of these ingredients.

You can be picky about ingredients when making your own products. Getting back to natural is as easy as growing fresh herbs in your garden. Even a simple container filled with a few plants will provide enough to harvest and create recipes. Fresh, handmade products feel different from store-bought and don't need a lot of chemical preservatives, since they are made in small batches. Take time to learn more about what your herbs can do for you and harness the healthy beauty from the garden.

This container is the perfect garden of herbs to place in an outdoor sanctuary space, a personal space in the garden where you can relax, read, or nap surrounded by the aroma of healing herbs. Comfortable seating and soothing colors are a must. What other things can you add to a patio space that bring comfort? Make time to pamper yourself and transform the herbs you grow into natural skin care remedies.

A bouquet of harvested chamomile and lavender is ready for natural beauty recipes.

Herbal profiles

Lavender 'Royal Velvet' (*Lavandula angustifolia* 'Royal Velvet'): Perennial. 'Royal Velvet' is a showy variety of English lavender with a mounding habit to fill in spaces with its silvery foliage. Deep-purple blooms are abundant in midsummer. If you want to up your lavender bud harvest from container gardens, grow French hybrid varieties (*Lavandula* x *intermedia*) such as 'Grosso', 'Provence', and 'Phenomenal'. These become more substantial than the English varieties, with a high stem count loaded with lavender buds for harvesting, and should be planted individually in their own pots, which can be top-dressed with decorative rock. In the cosmetic industry, these cultivars are the ones most used for their high oil content. From your own natural beauty garden, capture the essence by making an infusion of lavender buds in water to mist and soothe sunburned skin. Lavender flowers are popularly used as the main ingredient in soaps, bath salts, and lotions.

Parsley 'Extra Curled Dwarf' (*Petroselinum crispum* 'Extra Curled Dwarf'): Biennial, typically grown as an annual because it will go to seed and stop leaf production after the second year of growth. This dwarf form tucks nicely into containers. The beautiful deep-green, frilly leaves add fabulous texture in mixed plantings and can be used like a lacy collar along the edge of planters. The plants stay attractive all season and may overwinter as an evergreen in mild climates. A lesser-known attribute of parsley is that it is a super skin healer. An infusion into water or witch hazel is very mild and helps calm red, irritated skin or soothe a bout of acne. Saturate a cotton ball with freshly brewed parsley tea and gently smooth it over irritated skin.

Scented geranium, rose (*Pelargonium graveolens*): Tender shrubby plant. Protect from frost or bring indoors in the winter. Well suited for container gardens, this pelargonium can grow up to 3 feet (0.9 m) tall. Trimming green stems down to new leaf growth will encourage bushiness. There are many varieties of scented-leaf geraniums, but only a few are suitable for skin care and rose geranium is the best. In skin lotions and massage oils, its tonic qualities are highly valued for irritated skin. Its astringent properties help to cleanse away oils and grime from the skin to refresh and renew it. Infuse purified water with the leaves, strain them off, and pour the scented water into a mister bottle. Take it on your next hiking or camping trip to use for sunburns; it is also a natural insect repellant.

Scented geranium 'Lady Plymouth Variegated' (*Pelargonium* 'Lady Plymouth Variegated'): Tender shrubby plant. Protect from frost or bring indoors in the winter. This is a variegated-leaf form of rose-scented geranium. The color will stay bold and distinct when the plant is grown in full sun. Shady conditions make scented geraniums leggy and messy, and the variegation is unstable and may fade to a plain green leaf. 'Lady Plymouth

Variegated' is a beautiful ornamental, fragrant addition to a container garden. It stays bushy and full and adds color with its creamy variegation, plus the finely cut leaves lend an elegant lacy look. The leaves are highly scented and can be used the same way you use the green-leaf variety of rose geranium.

Enjoy an Herbal Foot Soak

Remember "time out" when you were a kid? It was all about taking the time to stop and get in a better mood or adjust your attitude. As adults, we probably don't take enough time outs. Busy days grab all the time we have, cell phones ringing, electronic calendars dinging; it can all be a bit much. Take a spa moment with herbs you grow in the garden. Here is a quick remedy to get you off your feet. Make this herbal foot soak blend so it's ready when time allows you to slow down.

1 cup (115 g) coarse sea salt

½ cup (58 g) baking soda

⅓ cup (9.5 g) dried scented geranium leaves

¼ cup (7 g) dried lavender buds

Crush the herbs into the sea salt and baking soda until they're in small pieces. Stir well and store in a tightly sealed glass jar.

Now do it! Grab a towel, a tub that will fit both your feet comfortably, this herb mix, a tablespoon, and enough warm water to cover your feet to the ankles. Dissolve 2 tablespoons (12 g) of the herbal mixture into the water. Sit down, relax, and soak your feet for a minimum of 15 minutes to enjoy the full benefits of the aromatic herbs. Ahhhh!

Herbs from the garden provide healthy choices for environmentally friendly housework. The list of helpful herbs for eco-friendly cleaning is long. Mint is a natural pesticide, lavender an air deodorizer, oregano an antimicrobial. Bay leaves placed in the pantry repel moths and other insects. Rosemary and sage contain camphor, which is a disinfectant, and thyme is filled with essential oils that have antibacterial and antifungal properties. You can capture these powerful components to make natural, fragrant household cleaners by adding dried and fresh herbs to ordinary ingredients from the kitchen, such as baking soda and white vinegar.

Create a tabletop garden of cleansing herbs early in the growing season. Use the herbs for spring cleaning and lightly trim fresh plants throughout the season as needed for use. When the plants outgrow their small pots in midsummer, cut them back (saving the cuttings and hanging them to dry to make herbal cleaning products) and transplant them to larger pots outside or directly into the garden.

Eco-friendly cleaner made with vinegar infused with rosemary and sage.

Herbal profiles

Lavender 'Munstead' (*Lavandula angustifolia* 'Munstead'): Perennial. The cultivar 'Munstead' is one of the hardier English lavenders, and the bloom type is the classic elongated lavender stem. Lavender is a charming container companion in planters with thyme and rosemary. They are all native to the Mediterranean region and are low water-use plants that do not need nutritious soil. Use the flower buds in household cleaning formulas—their oils are cleansing and aromatic to deodorize while cleaning away bacteria and germs.

Lemon verbena (*Aloysia citrodora*): Woody shrub, half-hardy in most areas as it is typically killed by heavy frost. Lemon verbena can grow large in a container and makes a nice backdrop to smaller-size pottery in arrangements. In the summer, cut off flowers to encourage lots of leafy growth. As cold weather sets in or before freezing weather in your area, harvest all the leaves and dry them. The fresh citrus aroma is the best of any

lemon-scented herb you can grow. The limonene in the leaves, also a component of lemons, is known for its solvent qualities to help clean grease and grime. Use the leaves fresh or dried to add an invigorating fragrance to natural homemade cleaning products.

Rosemary 'Arp' (*Rosmarinus officinalis* 'Arp'): Perennial. 'Arp' is one of the hardier varieties of upright rosemary and will take mild frosts in the late fall and early winter. The deep green leaves and upright growing habit make this a charming tall accent in the middle or back of a large container. Planted alone or with other herbs that also appreciate well-draining soil, rosemary can thrive for years. Essential oil in the aromatic leaves is highly antibacterial and disinfecting. Strip the fresh leaves from the woody stems and steep them in white household vinegar for a natural glass and counter cleaner.

Sage 'Berggarten' (*Salvia officinalis* 'Berggarten'): Perennial. 'Berggarten' sage has one of the largest leaves of the typical garden sages. The leaves are a soft gray that adds a shimmer of silver to container gardens. Easy to maintain, it can be trimmed to stay within its boundaries in mixed planters. In areas of cold winter, sage will go dormant. In the spring, tiny leaves emerge up and down the woody stems, and the plant fills out quickly as the weather warms. If it is planted by itself, a light seasonal pruning is usually needed to keep it from getting too woody and leggy. Sage is very similar to rosemary and thyme in its cleansing qualities. These Mediterranean herbs all have essential oils with antibacterial and antifungal components.

Scented geranium, peppermint (*Pelargonium tomentosum*): Tender perennial. This scented-leaf pelargonium has large, velvety leaves. The bold leaf adds standout texture in mixed plantings. For an extra kick of color in containers, look for the cultivar 'Chocolate Mint', which has a dark chocolate-colored splotch in the center of the big fuzzy leaf. In the heat of summer, this pelargonium will grow fast and can get long and leggy. Sometimes in a container the rangy growth is attractive; it will spill over the sides and weave around other plants. If you prefer the plant to stay bushy and tidy, selectively remove green stems just above a leaf. The essential oils in the leaves, which are used fresh and dried, are cleansing and help remove grime and leave a fresh scent.

Thyme, English (*Thymus vulgaris*): Perennial. Thyme is a mounding plant that stays compact in container gardens. You can start thyme from seed or, if you are impatient, find the plants in most garden stores. Because of its tidy habit, it fills in nicely around other plants and does well in mixed containers. English thyme is a common variety, dependable and very easy to grow. The leaves are saturated with aroma and used in household cleaning products to remove grime and grease. It also has antibacterial properties, making it one of the more potent herbs you can use.

Thyme 'Silver Posie' (*Thymus vulgaris* 'Silver Posie'): Perennial. This variety of thyme is both useful and ornamental; it has all the wonderful cleansing attributes of English thyme, with a bonus. The bright silver, variegated leaves create a slow-growing mound that highlights the edges of containers. In peak season, when grown in the full sun, it looks almost completely white. Cut a small bundle of stems, lay them in even lengths, and create a narrow bundle. Wrap kitchen twine around them to create small smudge sticks. Allow to dry completely. To use, burn slowly like incense to freshen and cleanse stale air inside the home.

> **OTHER CONTAINER HERBS FOR HOME CARE**
> basil
> lemongrass
> oregano
> peppermint

NATURAL SOFT SCRUB HERBAL CLEANSER

Harvest leaves and stems from your container garden to make this fragrant, mild scrub to clean sink, tubs, countertops, and glass stovetops.

> ¾ cup (173 g) baking soda
> ¼ cup (7 g) dried herb leaves, any of those in this section in combination or alone
> ¼ cup (80 g) unscented castile soap
> 1 tablespoon (15 ml) water

Combine the baking soda and herbs. Pulse the mix a few times in a spice or coffee bean grinder to break down the herbs into the soda. You don't want it to be a fine dusty powder, so it may take only a few pulses. Store the dry mix in a glass jar. To use: Add the dry herb mix to a small bowl, with the castile soap and water. Stir until it is a well-mixed paste. If it is too runny, add more soda and herbs; if too thick, add more water. You want it to be a spreadable consistency like toothpaste. You can add a few drops of rosemary or peppermint essential oil to ramp up the aroma and cleaning qualities.

CHAPTER 8

HERBS FOR POLLINATORS

PLANTS TO ATTRACT AND SUPPORT HUMMINGBIRDS, BUTTERFLIES, AND BEES

Attracting pollinators is essential for a healthy garden. Herbs top the list of pollinator-attracting plants. Colors send out a signal that there is something worth investigating there, and the aroma pulls in pollinators. The plants provide essential oils, pollen, and nectar—valuable nutrition for hummingbirds, bees, butterflies, and other insects that pollinate our gardens and keep them growing.

Colorful with summer abundance, a container garden filled with herbs to attract pollinators should be placed where it will be most beneficial. Target areas where the insects these herbs attract can also pollinate food gardens. Use multiple containers to create a pollinator buffet. Once the diners arrive, nature does the work to help make a healthy, productive garden.

Who doesn't stop to observe the mesmerizing flight of a hummingbird? I know I am always fascinated by the fast flittering from flower to flower and the beauty of the birds themselves. Gardens are made better by the activity, as hummingbirds are important pollinators.

Hummingbird-attracting gardens don't have to be large; even in small spaces, one or two plants will draw in hummingbirds like a magnet. The bigger and more dramatic the grouping of plants and containers, the more the color and aroma will draw in these flying jewels. Most attractive to hummingbirds are the tubular-shaped flowers of herbs such as pineapple sage, anise hyssop, and bee balm. Plant annual climbing nasturtiums around the edges of the containers for extra summer color.

Adding feeders to your hummingbird garden makes the buffet irresistible. Hummingbirds enjoy drinking from, playing in, and bathing in water misters and large water droplets, so provide small birdbaths to meet that need. Delve into the hummingbird species that frequent your area—whether year-round residents or seasonal migrants—and learn how to cater to them. You will be rewarded for your efforts.

Herbal profiles

Anise hyssop 'Blue Fortune' (*Agastache* 'Blue Fortune'): Perennial. Hardy and dependable. Agastache 'Blue Fortune' begins to bloom in early summer and continues to produce a beautiful flower spike all season long. The foliage emits a heavy licorice or mint-like aroma in the heat of the day. Other varieties of *Agastache* to add color and fragrance to a container garden include 'Blue Boa', 'Golden Jubilee', and 'Black Adder'.

Bee balm (*Monarda* spp.): Perennial. Blooms in midsummer with vibrant colors. The long, tubular flowers that accommodate the long beaks of hummingbirds make bee balm a must-have in this themed garden. The leaves are also highly fragrant, adding to their attraction. *Monarda* can be prone to powdery mildew when packed into mixed container gardens, so use cultivars that show resistance to this disease. These include 'Marshall's Delight', 'Raspberry Wine', and 'Violet Queen'. 'Petite Delight' is a newer dwarf variety that is good for container growing.

Mint 'Grapefruit' (*Mentha* x *piperita* 'Grapefruit'): Perennial. A tidy, low-growing plant, but still a rapid spreader like all mints. Works best if planted by itself in a container. The small lilac flowers produce heavily over a long season. Crinkly, soft leaves with a silvery, fuzzy texture make an interesting addition to a container grouping, but the abundant flowering habit of this mint is the real attraction. The fragrance is a mild citrus aroma and not as pungent as true peppermint.

Purple coneflower (*Echinacea purpurea*): Perennial. Easy to grow and not fussy. In containers, use compact-growing choices, such as 'Kim's Knee High', 'Pow Wow', 'Wild Berry', Pixie Meadowbrite™, and 'Ruby Star'. Leave the aged cones on the plants to attract other birds, such as goldfinches, which will eat the seeds.

Rosemary 'Barbeque' (*Rosmarinus officinalis* 'Barbeque'): Perennial. The cultivar 'Barbeque', so named because its long, woody stems are used as flavorful skewers for grilling meats and vegetables, grows tall and narrow, making it a good plant for height in mixed containers. For healthy root growth, start with well-draining pots at least 12 inches (30 cm) deep. All upright rosemary varieties look great in planters by themselves. Rosemary is especially useful for areas where hummingbirds overwinter. The flowers bloom in the winter and create a nectar source when other plants are out of season.

Sage, purple garden (*Salvia officinalis* 'Purpurea'): Perennial. This sage is a common garden type with soft purple leaves that mix well in containers and help offset lighter green colors. Hummingbirds find the lilac flower spikes that appear in midsummer very attractive. Once the flowers have faded, remove the spikes to allow the colorful leaves to show off again. Tidy this plant by removing older, woody stems as needed to keep it full and leafy.

Thyme 'Foxley' (*Thymus pulegioides* 'Foxley'): Perennial. Large, rounded green leaves with creamy variegation make this low-growing plant a good one to tuck around the base of taller plants in a container. The showy variegation can sometimes be heavy enough that the leaves look all cream colored. Throughout the summer, lilac-purple flowers cover the plant.

Color matters

Typically, we think of red as the color that attracts hummingbirds. It is indeed the color to which they are most sensitive as a signal that there's a nearby food source. But a mass of bright flowers of any color can also be alluring. In my garden, Anna's hummingbirds always visit my blooming rosemary hedge in the winter when it is covered with sky blue blooms. There are no red-blooming flowers that time of year. It is purely the aroma and the large mass of bloom from the rosemary that attracts them.

Interestingly, researchers who varied plants by nectar rather than flower found hummingbirds preferred certain nectars regardless of bloom color. Many herbs can offer both color and nectar to hummingbirds, and a container garden of such plants

Hummingbird Mint (*Agastache*) Blue Boa blooms abundantly and is a fragrant draw to bring hummingbirds to summer container gardens.

gives you the flexibility to draw activity to a specific location. Consider a place inside your home where you like to sit and look out at the garden. Add beautiful pottery filled with hummingbird-attracting herbs, then sit back and enjoy the show.

Hummingbird habitats

Before inviting birds, such as hummers, and beneficial insects, such as bees and others, to the garden, you need to understand the habitat in which they thrive and are safe. An ideal hummingbird habitat provides shelter, food, and water. Make your garden a well-rounded hummingbird haven by adding other food sources and plants that will shelter them.

Shelter in the garden near your herb containers will help keep hummingbirds close. Place your planted containers where there are larger shrubs and trees nearby—no more than 10 to 15 feet (3 to 4.6 m) away. The plants will give places for hummingbirds to easily perch and rest or hide from predators. Large, leafy shrubs or evergreens that are more than 12 feet (3.7 m) high will offer female hummingbirds a place to build their nests.

All hummingbirds are quite protective of their territory, and once they discover the haven you have created for them, they will remember and come back season after season.

> **OTHER HERBS FOR HUMMINGBIRD GARDENS**
> catmint
> nasturtiums
> pineapple sage
> roses

Use containers to place pollinator herb gardens where they will encourage insects to hang out. If your garden is limited and you are reserving valuable space for growing food, tuck containers amid edible gardens or urban beehives to help attract beneficial insects. Locate the containers away from home entries, walkways, and outdoor activity and dining spaces. Bees are more content to do their work in places away from a lot of human activity.

When planning to attract pollinator insects to the garden, study all of the things they need in order to stay. Consider plants that give something during all cycles of life. For example, herbs within the family Umbelliferae with flowers shaped like umbrellas, including fennel, dill, and parsley, are good host plants for the caterpillars of swallowtail butterflies. Bees are drawn to the flowers of high nectar-producing herbs such as chamomile lemon balm, bee balm, hyssop, mint, rosemary, sage, and savory.

Herbal profiles

Anise hyssop 'Blue Fortune' (*Agastache* 'Blue Fortune'): Hardy herbaceous perennial. This carefree, easy-to-grow plant draws in pollinators with its large spikes of blue blossoms. It will bloom for an extended period as the summer warms through fall. Deadhead flowers throughout the season to encourage repeated bloom. The leaves are aromatic and when crushed have a licorice-like fragrance, inspiring the common name anise hyssop.

PLANTS NEEDED
anise hyssop 'Blue Fortune'

PLANTS NEEDED

anise hyssop 'Blue Fortune'
catmint Junior Walker
lavender 'Anouk'
oregano 'Kent Beauty'
purple coneflower Pow Wow®

Herbs in the family Apiaceae, also called Umbelliferae, such as fennel and dill, host eastern swallowtail caterpillars.

Catmint, Junior Walker (*Nepeta* x *faassenii* 'Novanepjun'): Herbaceous perennial. The variety pictured at left is a low-growing form of a popular cultivar, 'Walker's Low' (which is also an excellent pollinator herb). Junior Walker's dwarf habit makes it ideal as a quick filler for containers. Use it to extend the season of purple color before and after lavender plant varieties fade. Its long bloom season starts in early summer and continues until cold weather in the fall.

Lavender 'Anouk' (*Lavandula stoechas* 'Anouk'): Woody perennial, tender in areas with below-freezing temperatures. The long bloom season of Spanish lavender (*Lavandula stoechas*) as compared to English lavender makes it a color performer in container gardens. It can grow big and abundant and, after a few seasons, may need its own pot. The silver foliage is also a highlight contrast when mixed into container gardens. Deep-purple blooms begin in the early summer and continue over many weeks.

◀ **Oregano 'Kent Beauty'** (*Origanum* 'Kent Beauty'): Low-growing perennial. This stunning and unique low-growing oregano is a bee magnet. The tissue paper–like flowers bloom all summer, and the rounded leaves have an aroma of oregano. It is not culinary oregano and not suitable for harvesting and cooking, but it is a flowering powerhouse. In containers, it will form a mat that when planted along the edges of the container drips down the sides to add texture.

Purple coneflower, PowWow® (*Echinacea purpurea* 'PAS702918'): Herbaceous perennial. Giant white coneflowers are showy and bloom for a long season. The plants are compact, making them an ideal container garden choice. Fuss-free when grown in a sunny, warm place, they bloom over most of the summer. Once the flower petals fade, leave the dark, spiny cones on the plants. Birds will pick at the seeds.

Fall-blooming black-eyed Susan extends the season of attraction.

Coming attractions

Numerous plants not necessarily classified as herbs also encourage pollinators to seek out your garden, and their bloom cycles let pollinators extend their stay beyond the peak season of most herbaceous herbs. To welcome spring arrivals, plant pansies and violas and spring bulbs, such as daffodil, crocus, and hyacinth. Add late summer- and fall-blooming plants, such as asters, goldenrod, and black-eyed Susan, to help fuel up migratory species for their journeys and local species for overwintering.

Here are a few insect-attracting shrubs and trees that can be planted in a large container alongside your herbs.

butterfly bush (*Buddleia*), Lo and Behold series
columnar apple and other dwarf fruit trees
crape myrtle (*Lagerstroemia*) 'Pocomoke'
hardy fuchsia
ninebark (*Physocarpus*) Tiny Wine®

OTHER HERBS FOR BEES AND BUTTERFLIES

bee balm
borage
chives
dill (especially attractive to ladybugs)
English lavender
lemon balm
parsley (good host plant for butterfly larvae)
rosemary
thyme

Make an Easy Bee Gathering Spot

Bees need a source of fresh water, and shallow dishes with places for them to perch and rest are ideal. This easy little bee gathering dish will support the bees that are attracted to your container herb gardens.

Clean the dish well and place it in an area near your container garden. You can make it part of a garden setting by turning empty planters upside down and elevating the dish as a part of an arrangement. Add marbles or stones and add fresh, clean water. Do not submerge the marbles, as you want them to act as a perch for the bees. Keep your bee bath healthy by changing the water daily. Do a full cleaning weekly or as needed to stop algae and other unhealthy growth in the water.

PART 3

PLANT AND TEND YOUR HERB CONTAINERS

To begin growing herbs in containers, often the easiest way is to head to the local garden shop or farmers market and purchase plant starts. Then the question becomes what kind of soil to grow them in and how to maintain and harvest them for health, visual appeal, and longevity. You also have the option of growing your own plants from seed or cuttings.

CHAPTER 9

FILLING AND MAINTAINING THE POTS

Bountiful container herb gardens depend on getting your pots off to a good start. Many herb plants are available at local garden stores and farmers markets. Soil mixes are also available at nurseries and garden shops, or you can make your own. Once you have your herbs planted, knowing how to care for and harvest from them will ensure that they live long, happy lives.

SHOPPING FOR HERBS

Shopping for plants will give you the quickest satisfaction for finished container gardens. It is also a space saver. If you do not have the time or space to start seeds or overwinter cuttings, your option is to look for healthy plants to add to your potted garden. Shop garden centers, farmers markets, and local plant sales to find variety and selection.

What to look for when buying potted herbs

Above the soil line of the pot, look for overall health, well-shaped plants, and, depending on the time of year, fresh new growth.

Then take a closer look: Inspect for bugs, weeds, and debris on the soil surface. Check the soil to make sure it is not dried out to the point that it has stressed the plant. You can usually tell if a plant has been watered regularly: The soil is somewhat loose, dark, and cool to the touch. If the soil surface is hard and crusted over or the soil ball is pulling away from the sides of the pot, don't purchase that plant.

Tap the pot: Thump on the side of the plastic nursery pot to get a feel for the root ball. A hollow sound usually means the root ball is bound up and there is no loose, healthy soil. The root ball may even slide easily out of the pot. If it is all a mass of roots that have formed in the shape of the pot, it is root-bound, and it will be a struggle to get the plant to do well.

Look at the bottom of the pot. A few stray roots coming out of the drain holes in the bottom of the pot are okay, but avoid large bound-up masses of roots clogging up the hole and growing outside of the pot.

For annual herbs, look for plants that are stocky and lush, with new buds and flowers that are not yet open.

The seasons of buying

Depending on where you live, there are typical times of the year when plants become available for purchase. To have well-rounded container gardens, you might need to leave spaces in mixed containers to add more plants or room for more pots as different herb varieties become available in season.

These **cool-season herbs** are typically available early in the spring and can be outdoors in colder weather:

- chervil
- chives
- cilantro
- dill
- fennel
- mint
- parsley
- violas

These **warm-season herbs** are usually available once the average temperature begins to rise:

- basil
- French tarragon
- lavender
- oregano
- rosemary
- sage
- stevia

FILLING THE POTS

Plants in containers cannot get water or nutrients on their own like plants in the ground can. That is why pottery that drains well and the right potting soil are vital to success with herb gardening in containers. When you get these two things right, your herb garden will grow well, be easy to maintain, and last a long time.

Drainage

If you remember one thing from this book, it should be the importance of good drainage in herb container gardening. A container must have a way for the water to drain away from the plants' root systems. Plants can literally drown if water has no place to go. It is a delicate balance in a container; the water needs to stay long enough for the roots to drink it in but drain away fast enough that they don't rot. Rocks in the bottom of the container do not make a well-draining container; the water still needs a way to escape from the pot. Depending on the size of the container, it should have a hole about ¾ inch (2 cm) in diameter every square foot (929 cm²) to give sufficient drainage. Check this and make sure you take care of it *before* you plant.

Soil

The soil in potted gardens is the messy but vital part of a beautiful container. Healthy, nourishing soil is typically a mix of organic materials, such as compost, peat moss, perlite, and sand.

How much soil do you need? Bagged soil is usually sold by the cubic foot. To figure out the volume of your container, use a bit of math to estimate the cubic feet of space. Basically, you multiply the individual container's length by its depth by its width. If you measure in inches, the resulting number is cubic inches. To get to cubic feet, divide that number by 1,728. (Note: This formula will be different if you use the metric system.) Multiply that result by how many containers you have, and add about 10 to 15 percent more, especially if your container is an odd shape.

ROCKS IN THE BOTTOM?

For many years we were told to place 1 or 2 inches (2.5 to 5 cm) of gravel in the bottom of a container to help with drainage. I did this and will admit to teaching it in classes many years ago, but I never noticed that it made a huge difference in helping the pot drain better.

Studies have shown that this method of planting containers is not a good idea. Potting soil is like a sponge and holds water until it is overly saturated, so water is not truly draining away through the gravel; it is building up at the bottom of the soggy soil layer. The extra layer of gravel is elevating that moisture layer higher into the root area of the plants. The boggy layer of soil at the bottom makes them more susceptible to root rot.

The hydrology of soil is a fascinating study; if it is your thing, explore more about this by researching "perched water table" or PWT. PWT is a zone of saturation above an unsaturated layer, usually made up of a different texture or substance.

Use this handy chart to figure out how much soil you need to fill a container. All calculations are an approximate measure of fluffy, dry soil volume. Add about 10 percent for unusual-shaped or odd-sized containers.

4-inch pot (10 cm) = 1 pint (0.5 L)
6-inch pot (15 cm) = 1 quart (1 L) or 0.03 cu. ft.
8-inch pot (20 cm) = 1 gallon (4 L) or 0.15 cu. ft.
10-inch pot (25 cm) = 3 gallons (11 L) or 0.46 cu. ft.
12-inch pot (30 cm) = 5 gallons (19 L) or 0.77 cu. ft.
16-inch pot (41 cm) = 10 gallons (38 L) or 1.5 cu. ft.
24-inch pot (61 cm) = 25 gallons (95 L) or 3.8 cu. ft.

Hanging baskets:
10-inch (25 cm) = 5.5 quarts (6 L) or 0.21 cu. ft.
12-inch (30 cm) = 7.9 quarts (8.4 L) or 0.3 cu. ft.

Window boxes:
24 inches (61 cm) long by 8 inches (20 cm) deep = 11.7 quarts (12.8 L) or 0.45 cu. ft.
30 inches (76 cm) long by 8 inches (20 cm) deep = 15.6 quarts (17.1 L) or 0.6 cu. ft.

Bagged soils

I find most container gardeners are limited on space and don't have room to mix up potting soil. Purchasing bagged products might be a better option, especially when the soil needs to go upstairs or through your home to access your herb gardening space.

Become a label reader. Many mixes on the market have significantly different ingredients. Make sure it is labeled as a potting soil and not a seed-starting mix, garden mix, topsoil, or straight compost formula. Bagged potting soils are typically made up of elements that provide good drainage and nutrients that plants confined in pots need. I look for good basic organic ingredients, such as compost, perlite or vermiculite, coir, and sphagnum peat moss.

I skip bagged products that have a lot of extra things, such as water-holding granules and time-release fertilizers. I prefer to add fertilizers and control the moisture content the old-fashioned way. At the time of purchase, the bag of soil should be light, based on the size of the bag. Heavy bags may have too much sand in the mix; or they could be saturated with water, making them smelly, anaerobic, and unhealthy to use.

There are many commercial potting soils you can use to grow herbs in containers.

Making your own container blends

The choice between bagged soils or bulk comes down to how many containers you need to fill and where you locate them. Mixing and hauling bulk bins of soil up to a rooftop garden is not for everybody, but if you are filling a large number of containers or several of substantial size, making bulk blends will be less costly. Another reason to go with your own mix is to get a custom blend; it seems all gardeners have their favorite brand or blend of ingredients.

Here are the ingredients of a typical soil mix:

- soil (commonly used as a base; purchase sterilized potting soil)
- well-rotted manure or compost (adds natural nutrients; fresh manure can burn plants and is likely to contain unwanted seeds and other contaminants)
- perlite (a volcanic glass, familiar as the little white specs in most pre-made mixes; creates air space, lightens overall weight, promotes drainage)
- vermiculite (aluminum-iron magnesium silicates, small shiny mica-like bits in soil mixes; improves aeration and moisture retention)
- sphagnum peat moss (holds moisture, increases soil acidity; valuable binder to help blend other ingredients)
- coir (brown fibers of coconut husks; a natural, sustainable substitute for peat moss; aids moisture retention, loosens texture of soil; more pH neutral than peat moss)
- sand (adds drainage to heavy mixes; purchase sterilized to avoid contaminants)

If making your own blend of soil is appealing and you have the space, locating the separate ingredients is the key. An easy option is to purchase bagged ingredients from a local garden store.

BASIC SOIL MIX FOR CONTAINERS

1 part bagged, sterilized garden soil

1 part peat moss or coir fiber

1 part perlite or vermiculite

1 part compost

Moisten all ingredients and mix well in a roomy container, such as a wheelbarrow or large tub. Feel the texture of the soil and adjust your mix. If it feels heavy and sticky, add a handful more of peat or coir. If it feels too light and needs more substance, add compost.

HOW TO MAINTAIN YOUR HERB-FILLED CONTAINERS

The low-maintenance nature of container growing makes this one of the easiest types of gardens to enjoy. There is no need for weeding, mowing, raking, or edging, just a few important tasks to keep your container garden healthy and looking its best.

Watering

Regular watering needs to be a part of a container garden routine. Potted plants cannot grow deep into the soil to find moisture; you are their only source of water in dry weather. When containers are dry, water the soil evenly until it is drenched and water is pooling on the surface. Avoid overwatering drought-tolerant herbs, such as rosemary or lavender. Also avoid watering in the evening, when water is less likely to evaporate from leaves, making plants susceptible to diseases, such powdery mildew.

Adjust your watering schedule based on the plants' needs, container type, and the weather. Clay pots can dry quickly in the heat of the summer, so check those often. In rainy periods, don't expect that rainfall is sufficient enough to water container gardens. Check them regularly—the canopy of plants may keep rainwater from getting to the soil. Also check sheltered herbs, such as those under eaves, to ensure they are getting enough moisture.

Fertilizing

Container plants have few sources for nutrients. The organic matter in potting soil quickly breaks down, so planters need fertilizer once in a while. A tightly planted window box, living wall, or hanging basket has less soil mass to hold nutrients and will need fertilizing more often.

Most herbs are grown for their foliage and not much for their blooms, so choose a balanced general-purpose natural or organic-based product. Garden centers have many types available, but for container gardens, I find that the easiest is a liquid fertilizer. Check the label of the product. Most formulations can be mixed in a watering can and poured over the foliage and the soil. Make sure to deeply water at the soil level to insure proper nutrient distribution to the root system.

Top-dressing

Give container gardens with an exposed surface area a finished look by top-dressing, or adding an extra element to the soil surface. The topdressing also helps retain moisture and protects the plants. During the summer, if outdoor containers have weeds, moss, or algae growing on the surface, a topdressing helps keep the surface clean. Light colored, decorative rock or beach glass on the soil surface also prevents too much moisture build up around the base of plants. Topdressing is especially useful for Mediterranean herbs such as rosemary, lavender, and sage, which prefer drier soils. Before adding any type of topdressing, remove weeds, debris, or pests.

Opposite: Peppermint is dramatically displayed in an olive jar–style terra cotta pot.

Terra cotta shards from broken pots are handy for top-dressing a potted herb.

Here are some popular topdressing materials:

- **Hazelnut shells:** Broken hazelnut shells are a by-product of the nut industry and are available in some areas as a bagged topdressing for garden beds. Use the shells on the surface of the soil to discourage slugs and pests and to retain moisture.
- **Decorative rocks, gravel, beach glass:** Herbs that do not like moisture around their base do well with a layer of rock. It especially helps prevent browning around the base of lavender.
- **Terra cotta pieces:** Save pieces of broken terra cotta pots to use as a topdressing. They have a natural look, retain moisture, and keep the surface of the soil tidy and weed-free.

REPLENISHING SOIL

Herbs thrive in loose, organic soil. The soil structure in containers changes from season to season and can benefit from being refreshed. Adding or removing plants revitalizes the soil by loosening and amending it. If you are just adding or replacing plants in an established planting, remove root debris and mix in a high-quality potting soil. Most have the right balance of compost and nutrition to refresh the soil structure.

Repotting

Container gardens are not meant to occupy the same pots forever. Most perennial herbs, as well as shrubs and trees, need repotting every few years. You will need to evaluate living walls, hanging baskets, narrow window boxes, and shallow containers for repotting every year. Some herbs will stay longer if planted in a container large enough to support their growth for an extended period. Consider repotting to be part of the ongoing maintenance of container gardening.

Roots do not stop growing. They circle around the inside of the container or grow through the drain holes. Routinely check pots for signs that the roots are bound up. The plants will look off-color and always seem to require water. If you notice while watering that the water rushes out quickly without saturating the root ball, you need to remove the plant and check the roots. A healthy root system allows water to percolate through the whole root ball and has some space to allow nutrients and air.

Opposite: A top-dressing of small, rounded pea gravel on herbal centerpieces prevents soil from spilling onto the tabletop.

Here are some other symptoms of root-bound pots:

- Weak growth or no growth in the plant's center may be accompanied by new growth at the edges of the container.
- Small roots are becoming matted on the soil surface or growing out of the drain holes in the bottom of the pot.

Herbs need to be repotted and potted up from time to time as they grow.

Do not repot or disturb plants in the middle of their active growing period or bloom season. In harsh winter areas, bring herbs inside for protection before frost and, if possible, keep them in the same pot and repot in the spring when warmer weather arrives.

To repot a container garden, carefully remove plants from their pots and get rid of all the old potting soil if it is weedy or mossy on the surface. If the soil is okay to reuse, add organic matter back into the soil by mixing 50–50 with a soil amendment, such as high-quality compost. Prune and clean up the roots and put the plant right back into its container, keeping it at the same soil level as in its previous pot. Root-bound perennials need to be removed. Divide them and carefully break up the root system before placing them back in the pot.

PESTS AND DISEASE

Every garden contends with pests and disease, and containers are no different. Preventive care is the best tactic. Do some simple housekeeping by removing dead leaves and debris that collects in the top of the pot. Keep an eye out for spots, bugs, or leaf color that seems off. Move problem pots away from healthy pots until you clean up and help nurture plants back to health.

Keep in mind that a certain number of pests are unavoidable but overwhelming numbers are preventable on some levels. Bugs and disease really love to attack unhealthy, neglected plants. The healthier you keep your garden, the more resilient it will be. Well cared for containers use the techniques of organic gardening: Observation, prevention, and quick action go a long way in avoiding the need for artificial intervention. Use the least invasive method to control problems if they arise. For instance, you can rinse most bugs off the plants with a good shot of water or pick them off to remove them from the scene entirely.

Many types of natural disease controls exist to help keep your plants healthy. Discoloration and disease are not uncommon in crowded, mixed containers. The first step to effectively treating such a problem is to identify it properly. Then, treat it with the least invasive method possible. The following is a simple solution for powdery mildew.

Treating Powdery Mildew

Powdery mildew can be a problem with some herbs. Bee balm and common garden sage are particularly prone to it when crowded with other plants in mixed containers. This simple spray disrupts the disease cycle and helps plants stay healthy.

> 1 quart (946 ml) water
>
> 1½ teaspoons (7 g) baking soda
>
> 1 teaspoon mild liquid soap, such as castile or liquid dish soap (not a heavy degreasing formula)

Combine all the ingredients in a jar and shake until the baking soda blends with the liquid. Pour some into a spray bottle. Use the spray outside on plants during cloudy weather or away from the hot sun. Mist the tops and undersides of leaves. This spray works best if you catch the problems before they are full blown. I typically do a preventive spray on my bee balm plants in early summer, when warmth and humidity can trigger the disease.

HARVESTING

When harvesting herbs for use in peak season, you don't want to lose the look of the container. Be strategic. How you harvest plants can make a difference in their appearance. First and foremost, keep containers healthy and abundant, so you have plenty of herbs to pick.

Harvest leaves and flowers in as whole a form as possible, and chop or crumble them just before use. Every time you break or cut a leaf, it releases flavor and aroma. If I am going to harvest numerous leafy stems, I prefer to water the plants well and lightly rinse the leaves *before* harvesting. Harvest, and gently shake herb bundles to remove pests and excess water.

When to harvest

If you are growing your herbs for a specific use, such as cooking, natural skin care, or tea making, you can harvest fresh leaves, flowers, and stems any time during the active growing season. For the best flavor, and especially when you are drying plants for storage in the off season, collect plants in the morning after the dew dries but before the heat of the day. The heat from the sun releases the essential oils from the plant, which is where the flavor and aroma are. You want as much of that in the leaves and flowers as possible. But if it's dinnertime and you need fresh herbs, it's okay to cut what you need right then.

The best flavor and fragrance of some herb leaves, such as basil, marjoram, and oregano, can be had just before the plant begins to flower. Mint leaves are more flavorful when picked on young green stems; if mint plants get woody, the leaves become rough and have less flavor. You can harvest tough evergreen perennials, such as rosemary, sage, thyme, and winter savory, throughout the year.

Flowers, such as roses and calendula, are best harvested just before the flower petals open. Pick them individually and avoid bruising them too much. Lay them flat on trays or screens to dry.

When rosemary gets large in its container garden, tidy and cut long stems to bundle and dry. Use dried leaves in tea and seasoning mixes.

Harvest lavender stems just as flower buds are showing vibrant purple color but before the petals open. Depending on the season of the plant, harvest herbs that are fading or will not overwinter in containers. Store dried herbs in glass jars and label your harvest.

Place delicate herbs that do not hang well, such as chamomile and calendula flowers, on flat drying screens

A handy way to dry herbs is to bundle them and hang them upside down on a dowel laundry drying rack.

Planning for succession

Cutting and harvesting your herbal bounty may make it difficult to keep a container attractive. Herb containers need to strike a balance between beauty and function. The key is to plan for a succession of herbs through the year.

In the planning stages, think through all the seasons of your container herb garden. When will you be harvesting, and will you remove all of a plant or can you lightly trim it to get what you need? If space allows, don't plant all your heavily harvested herbs, such as chives, basil, and parsley, in the same pot. Add in other plants that have substance or can fill in around the harvested plants. The goal is to avoid creating too many gaps with your harvesting, and, when you do create a gap, to have a plan for what will fill it. A combination of perennial herbs and faster-growing annuals is a good way to minimize the seasonal impact of harvesting.

If you harvest a whole basil plant, replace it with a fresh crop of basil or tuck in an annual herb, such as calendula, 'Lemon Gem' marigolds, or violas. Add other seasonal companion plants to herbs, such as greens and edible flowers. Large, colorful leaf lettuce can fill a space in the spring. As the weather gets warmer, harvest the lettuce and add a basil plant in the gap. Combine calendula and nasturtiums around the edges of upright rosemary to add color. When the calendula flowers fade, the nasturtiums will keep going through the rest of the summer. When both flowering annuals are done, the rosemary will fill in the pot and bloom in late winter to early spring.

Get to know which seasons are best for which herbs and give them a companion during those parts of the year when herbs are not looking their best. A good example of a purple-themed pot is to plant English lavender and catmint together. The lavender is at its best in midsummer, and, after the lavender flowers fade or are harvested, the catmint will continue a mass of showy lavender-colored blooms until fall.

The bright orange blooms of calendula look beautiful in a mixed herb pot.

Tips to Keep Containers Tidy but Still Showy

- Seasoning or a cup of tea only require a small amount of most herbs. Harvest from the back or middle of plants rather than from the leaves in the front of the container.
- You might need just the flower (calendula or chamomile), but cut the whole stem so that there are no unsightly chopped stems poking out.
- Nip off buds and flowers of basil, mint, oregano, and similar plants to keep them leafy and lush.
- Renew cut plants, such as mint and oregano, that flower later in the season. Look for new growth at the base of the plant, and cut down all the old leaves and stems. New growth will renew the plant and flush it out full again if it is not too late in the season.

FILL-INS TO PLANT AROUND SEASONAL HERBS

Spring fillers to tuck around dormant perennial herbs:

chives
cilantro
curly parsley
leaf lettuces
pansies
spinach
sweet peas
violets

Summer fillers to plug gaps left by early harvest herbs:

basil
bush beans (look for unique colored varieties, such as 'Royalty Purple Pod')
fennel
kale
lemongrass
nasturtiums
shiso
sunflowers (dwarf)

YOUR CONTAINER GARDEN CALENDAR

Use this list as a reference for the seasonal tasks of maintaining a container garden.

Spring

☐ Evaluate pots for the season. Remove dead plants and clean off soil.

☐ Check for emerging perennials that have expanded out of their boundaries.

☐ Transplant, divide, and thin out plants as needed. Move herbs in small containers up to larger ones.

☐ Clean and disinfect empty pots to ready them for planting.

☐ Clean planted containers; remove any dead or old growth.

☐ Water with liquid fertilizer and refresh soil as needed.

☐ Sow seeds of annuals.

Summer

☐ Remove weeds growing in the soil and tidy plants as needed.

☐ Keep a regular watering routine as summer weather gets hotter.

☐ Fertilize with an organic liquid feed; check the label directions for correct timing and dosage.

☐ Harvest herbs any time they are ready and throughout the active growing season to keep leafy herbs lush and full.

Fall clean out. Dead basil and other annual herbs are removed and added to the compost pile. Tender lemongrass is removed and potted to a smaller container to bring indoors and protect over winter.

Spring refresh. Be sure to remove all dead plants and root systems from the container. Stir up soil to loosen and check for debris. Replenish with new soil if needed.

Before planting, clean and disinfect empty pots.

Fall

- ☐ Collect, dry, and place seeds in small paper envelopes labeled with the variety and date collected.
- ☐ Clean pots and cut back old growth.
- ☐ Protect tender plants from frost.
- ☐ Prepare pottery to be protected from harsh winter weather. Clean and store any pot that will be empty over winter. Check the drainage of any pot left in the garden to make sure water does not freeze inside it.
- ☐ Take cuttings of plants you want to overwinter indoors.
- ☐ Trim any excess growth and water small potted plants well before bringing them indoors.
- ☐ Harvest all annuals and tender plants that will not make it through the winter.
- ☐ Dry and preserve herbs for winter use.

Winter

- ☐ If you live in an area with mild winter, you can leave containers as is or move them to a sheltered spot for protection during weather extremes. Potted plants on balconies, rooftops, and decks that are exposed to harsh winds should be moved close to a building or into an area where they can be protected.
- ☐ Protect living walls with a frost blanket.
- ☐ Water plants as needed until the soil in the container is frozen. Do not water frozen pots, because the plants are unable to absorb the water.
- ☐ Group containers close together for extra protection.
- ☐ Clean up storm damage. If a pot breaks over winter, remove the plants and place them in another pot or wrap them in burlap to protect the roots until you can get them back into another container in the spring.

Curly parsley is a tough perennial that will withstand light frosts and still be harvestable in mild winters.

Herbs tucked in to overwinter in a greenhouse.

CHAPTER 10

GROWING NEW HERBS: PROPAGATION

Propagation, simply put, is a way to get more plants. You might already be familiar with the basic methods: seeds, division, cuttings, and layering. How you propagate for container gardens is no different, but how you want your planted creations to look might dictate the best method. Larger rooted cuttings and divided plants give quick gratification. A planned supply of fillers and blooming annuals through the season is easy to start from seed. Propagating herbs yourself is an economical way to maintain an abundance of choices and a steady supply of plants to keep potted gardens full while you continue to harvest.

HERB PROPAGATION CHECKLIST BY PLANT

There are usually multiple ways to get new plants, but the best propagation technique for one plant may not work well for another. This list gives you the easiest options for each plant listed.

Herb	Type	Best propagation method
aloe vera	perennial succulent	division
anise hyssop	perennial	cuttings, division
artemisia	perennial	cuttings, seed
basil	annual	seed
bay laurel	woody shrub	cuttings
bee balm	perennial	cuttings, division, layering
borage	annual	seed
catmint	perennial	cuttings, division
chamomile	perennial	seed, division
chervil	annual	seed
chives	perennial	seed, division
cilantro	annual	seed
dill	annual	seed
elderberry	woody shrub	cuttings, division
fennel	annual	seed
French tarragon	perennial	cuttings, division
garlic chives	perennial	seed, division
ginkgo	woody tree	cuttings, seed
lavender	perennial	cuttings, seed, layering
lemon balm	perennial	cuttings, seed, division
lemongrass	perennial	cuttings, division
lemon verbena	woody shrub	cuttings
lovage	perennial	seed
marjoram	perennial	cuttings, seed, division
mints	perennial	cuttings, division, layering
nasturtium	annual	seed
oregano	perennial	cuttings, division
parsley	biennial	seed
patchouli	tender perennial	cuttings
pot marigold (calendula)	annual	seed
purple coneflower	perennial	division
rose	woody shrub	cuttings, seed
rosemary	perennial	cuttings, layering
saffron	bulb	bulb
sage	perennial	cuttings, seed, layering
santolina	perennial	cuttings, layering
savory	perennial	cuttings, seed, layering
scented geranium	tender perennial	cuttings
shiso	annual	seed
signet marigold, Gem series	annual	seed
stevia	annual	seed
tea plant	woody shrub	cuttings, seed
thyme	perennial	seed, division

GROWING FROM SEED

Growing herbs from seed offers economy and diversity. A packet of seeds is usually a few dollars and will start numerous plants. Seed suppliers offer varieties that may be hard to find in local garden centers, and a search through a favorite seed catalog expands your container garden choices from ordinary to exciting. Look for colorful-leaf varieties of basil, shiso, the Gem series of signet marigolds, and unusual edibles to fill in around perennial herbs such as mustard greens, kale, and leaf lettuces.

In the garden marketplace, common herbs are easy to find, but you might want to grow a specific cultivar. One example is chives. You can typically find seeds, starts, and a neighbor who will give you starts of common chives, but garlic chives are harder to locate. The flavor is worth searching for, and you will find it from seed suppliers.

A downside to seeds is they are not ready for immediate planting. Timing and planning are helpful. You need to start seeds and continue to transplant seedlings as they grow until they get large enough to fill a decorative container nicely. Annual herbs are good choices to start from seed for quick results.

This dill seed is ready for harvest.

Collecting herb seeds

Collecting seeds is not magic; it is a ritual of gardening as old as civilization and another aspect of gardening to learn. Generations have passed along stories of seeds tucked in a saddlebag for the long journey West or cherished heirlooms brought by immigrants to remind them of home. To start saving seed for the next season of your own garden, get to know more about the plants from which you want to obtain seed. Understanding their cultural attributes is the first step.

You need to know a plant's life cycle, whether it is an annual, biennial, or perennial, and how long it takes for it to grow from seed in the ground to producing seed of its own. Plants that grow from seed to seed in one season are annuals. Biennials produce seed in their second season. Perennials typically produce their seeds in the second season of growth and beyond. Depending on the variety, most cultivated, herbaceous seed-bearing perennials mature their seed within two to three weeks after the bloom dies. Ever-bloomers continually produce seed as they complete a cycle of fresh bloom.

Collect seeds in small envelopes and label them with the variety and date collected.

EASY HERBS FOR SEED COLLECTION

Basil: Allow some of the flowers to mature. As the petals drop, the stalks will begin to dry and turn brown. Split open the flower base. If the seed is black, it is ready to harvest.

Chives: As soon as the purple flower begins to dry out on the plant, small black seeds will appear deep within the blossom's base. Wind or rain will quickly disperse the seeds, so keep an eye on them.

Dill and fennel: Cut the flower stalks just before seeds begin to ripen and turn a tan color. Place a small paper bag over the flower heads and cinch it closed around the stems with a twist tie, then hang it upside down to dry further. As the seeds ripen, they will drop to the bottom of the bag.

Dried chive seeds embedded in a flower left on the plant to dry.

The next step is learning where the plant produces the seed and what it looks like. Most plants produce seeds at the base of their flowers. If left on the plant to ripen and dry, those seeds will become viable for the next season. Some clusters of seeds, such as peas in a pod, are large and distinct, while others, such as chive seeds, are small and embedded at the base of a flower in the dried petals.

A seed collector quickly learns about survival in the garden. Self-sowing plants have a genetic determination to reproduce, so they have a point of dispersing seed at the opportune moment. Wind, rain, temperature, insect visits, birds—all are ways nature signals plants to let loose with seeds. A poppy pod closed one day may burst open the next and spill the seeds before you can collect them. Knowing and watching bloom and seed pod formation will help you not miss out on collecting the seed before it falls to the ground or gets eaten by birds.

Recognition of where the seed is formed and what it looks like will also help you determine when it will be ripe to collect. Seeds collected too early will not be viable. Most seeds are ready when they turn a darker color and dry out, a condition of dormancy until the time and place are suitable to grow a new plant.

Many common herb, vegetable, and flower seeds germinate as soon as they are planted under moist conditions and the seed begins to absorb water. Timing and temperature also factor into the process. This is why we need to store seed in a cool, dark, dry place. Collect seeds and allow them to dry, and then place them in envelopes or paper bags labeled with the variety and the date. If you use plastic bags or jars, make sure there is no buildup of moisture inside.

Starting herbs from seed

Some seeds need to be started indoors to get a jump on the season, and some are fine planted directly in your container outdoors. Getting a head start on seeds means that plants are ready to go outside when the season begins. Check the germination rate. Herbs that need a long growing season will be the ones you want to start ahead of time.

Starting seeds in succession indoors through the spring is a great way to have plants ready to fill containers as the season warms up. It's all about timing: Most seed packets will tell you when you can plant outside and ideal time frames to start seeds. Figure out when you want to put seedlings in the garden, and then count back the appropriate number of weeks based on the plant.

Success at seed starting depends on having the right supplies and growing environment.

Seed trays and/or pots: Almost any container will work to start seed as long as it will hold soil, drain well, and give room for emerging seedlings. Plant flats, peat pots, egg cartons, newspaper, repurposed plastic containers, and terra cotta pots are all good options. Repurposed plastic containers must be clean; give them a good rinse with a

USEFUL PEAT PELLETS

Peat pellets are seed-starting ingredients compressed into dry, flat discs. When placed in water, they expand to create mini soil pots. These are handy if you have limited space to start seeds or just want to start a few seeds without having to purchase a whole bag of soil. Place the expanded discs in a tray and drop a few seeds in each. Keep the mini pots moist and the air around them humid. Once seedlings emerge, add light for strong, healthy growth. When tiny roots begin to come out the sides of the pot and seedlings are 3 to 6 inches (8 to 15 cm) tall, plant them in containers, mini pot and all. No need to remove the fabric holding the pot together or disturb the roots.

Peat pellets are an easy way to start seeds for many herbs, including this basil.

weak bleach solution (2 tablespoons [30 ml] of bleach to 1 gallon [3.8 L] of water) and then rinse well with water.

Soil: Seed-starting mixes are blends of organic ingredients that keep the soil lightweight and water absorbent. Light, fluffy soil is best to give seeds air space and the moisture to get good germination. You don't want dense, nutrient-rich soil or heavy garden soil. Loosely fill your seed-starting container with moistened soil to just below the rim; don't pack it down too tightly. Then, plant your seeds. If your seed packet doesn't recommend a seed depth, a good rule of thumb is about four times as deep as the size of the seed.

Add a clear dome over plants to trap humidity and allow in light.

Water: Constant humidity is vital in the first weeks of seed germination. Seeds and the soil medium need to be kept consistently moist but not soggy. Create a mini greenhouse around newly planted seeds with a clear plastic cover or dome. The cover will keep humidity in so soil and seeds don't dry out. Once you see seedlings rise above the soil surface, you can remove the cover; but water them consistently.

Light: Overhead and direct light is vital after germination. Some seeds germinate in the dark, but once they emerge, they need sunlight in order to grow stocky and healthy. The bright light from a window is okay, but a grow light turned on for ten to twelve hours a day is better at keeping plants from becoming weak and stretching for light.

Warmth: Most herb and vegetable seeds need temperatures of 65°F to 75°F (18°C to 24°C) to germinate. Applying heat from beneath to keep the soil warm is ideal. Use seed germination heat mats or place the seed tray on top of an appliance that generates warmth. Once the seed has sprouted, you can remove it from the heat source.

Once seeds germinate and produce their first set of true leaves (the leaves that grow right after the first two leaves), you will need to transplant the seedlings to a larger pot to allow room to grow healthy roots. Don't let plants get root-bound in seedling trays or you will set them up for failure. Fill a 4-inch (10 cm) pot with moist potting soil and push a hole in the center of the pot to accommodate the new plant. Carefully remove plants from the tray and break apart any grouping of seedlings into individual plants. Gently place one per pot and firm the soil around the plant. Continue caring for the tiny plants by keeping them moist and in direct light. Allow plants to get vigorous and fill in with more leaves before moving them to their container garden home.

Before plants can be moved outside, they need to be hardened off. This is a critical step to set up new plants to withstand an outside environment. The idea is to get the plants used to sun and wind exposure gradually. Take young plants outside during the warmth of the day and place them in a shady place. Bring them back indoors at night to protect them from a dip in nighttime temperatures. Do this indoor-outdoor routine daily for two or three days. After that, change the location to a sunny spot and continue bringing them in and out for another three or four days. Make sure to keep them well watered during the hardening-off process. After a week, the plants should be adjusted to outside elements and can be put in containers.

CHAMOMILE TEA FOR FUNGAL ISSUES

Damping off is a common disease that destroys new seedlings. It occurs, of course, in damp conditions and can cause tiny plants to collapse. A fresh brew of chamomile tea is a simple, nontoxic solution. Add one chamomile tea bag or approximately 2 teaspoons of dried chamomile flowers to 3 cups (705 ml) of hot water and allow it to steep at least 6 hours or overnight. Remove the tea bag or strain flowers from the water and put the infusion in a spray bottle. Spray small seedlings as soon as they emerge through the soil and continue regular misting with the chamomile-infused water until they are strong little plants with their true leaves.

GROWING FROM CUTTINGS

Some herbs, such as French tarragon, do not produce viable seed and must be grown from cuttings or division. In learning to propagate herbs from cuttings, I started with easy ones, such as mint and scented geraniums. Both of these start fast and don't fuss much about the conditions in which you grow them. Experimenting with them is a good way to see how nature works and what part of the plant will sprout roots.

Cuttings involve snipping off a piece of the parent plant and putting it into soil or water to grow roots. You can easily identify soft stem or softwood cuttings—typically the quickest to start—as the new growth of woody herbs and shrubs that has yet to harden; scented geraniums, mint, lavender, lemon verbena, and rosemary are all good examples. Semi-hardwood and hardwood cuttings from woody shrubs and trees, such as bay laurel and elderberry, may take up to a year to root. You will need to have patience and plan to wait for plants to get large enough to use in a container.

Stem cutting is the best way to get a clone of the plant. The flavor, leaves, and flower color may vary from plant to plant, even in the same varieties. Cutting can assure that you get what you like about the parent plant. Have you ever noticed how common peppermint can be sweeter or a little more medicinal tasting? Sometimes peppermint tastes different based on the cross-pollination of its seed. When you find a tasty mint, keep its memorable flavor by take a cutting instead of saving its seed.

These scented geranium cuttings are ready to grow to a larger size for filler in aromatic container plantings.

Soil rooting

Most often, cuttings are rooted in soil. Here's how to do it.

Step 1: Prepare seed-starting cell trays or individual pots by filling them with a moist, lightweight soilless mix (the same as a seed-starting type of mix). The size of the pot or tray depends on the size and type of cuttings. Most cuttings do well with a 4-inch (10 cm) pot's size and depth. Use a dibbler or the end of a pencil to poke holes in the soil where you will place each cutting.

**TIPS FOR
CODDLING CUTTINGS**

Never let cuttings dry out. Keep them well watered, and create a humid, evenly moist environment by using a transparent covering.

Keep freshly planted cuttings in good light but out of direct, hot sun (photosynthesis still needs to happen).

Use heat mats to keep the soil consistently warm. Keep the air temperature between 55°F and 75°F (13°C to 24°C) and encourage root production.

Step 2: Just below a leaf node (where a new leaf emerges from the stem), cut a 2- to 6-inch (5 to 15 cm) piece of stem that includes a terminal bud (the bud at the very end of the stem). Remove lower leaves to clear the 2 or 3 inches (5 to 8 cm) of stem that will go into the soil.

Step 3: Place the end of the cutting into the soil and firm the soil around it to make good contact. Put the tray or pot in a warm area and cover it with a clear plastic bag or a rigid plastic greenhouse to keep the cuttings moist while letting in light. Make sure the covering is large enough that the leaves of the cutting do not touch the plastic.

Step 4: Test for root growth by gently tugging at the stem. If there is resistance, roots are forming. Watch for new leaf growth. Timing varies by variety of plant, but most cuttings form roots in about three weeks. Once the plants have a well-developed root system and are growing healthy new leaves, they are ready to place in containers.

Water rooting

You can root some herbs, especially moisture-loving ones such as those in the mint family, in water. Here's how to do it.

Step 1: Cut a nonflowering, leafy stem about 6 inches (15 cm) long. Strip off lower leaves to clear 2 or 3 inches (5 to 8 cm) of stem to go in the water.

Step 2: Put the cutting in a clear glass jar filled with water and place it in full, bright light. Depending on the size of the jar, include only two or three cuttings. Don't crowd them.

Step 3: Keep the water clean and remove any dead leaves or other debris. Keep the cuttings in the jar until you see a good, healthy mass of roots growing in the water.

A cutting of peppermint grows roots in a glass jar.

Step 4: Once a healthy, strong mass of roots has formed, gently remove the cuttings from the water and plant them in moist potting soil. Allow them to establish well there before planting them in container gardens.

EASY HERBS TO START FROM CUTTINGS

Scented geraniums: Can be cut and propagated any time of year, but the stems with a new flush of growth in the spring are ideal. If you are in a climate where you cannot overwinter the plants outdoors, take cuttings to bring indoors and grow new plants over the winter.

Lemon verbena: Take cuttings in early to mid summer when there is lush new growth but before stems become woody.

Lavender, rosemary, and sage: The best time of year to take cuttings is in the spring or early summer when the plants are actively growing. Avoid the woody part of the stems; look for tender growth with three or more nodes (the place where new leaves emerge on the stems.)

GROWING BY DIVISION

Most herbs with fleshy, clump-forming roots are easy to propagate by division. Simply break a mature, healthy plant into parts without damaging the roots or base to create smaller, individual plants. The best candidates are perennial herbs that die back and go dormant in the winter. Once they emerge in the spring, ready your empty pots and potting soil; dig up the plants; and use sharp, clean tools to divide them.

> **HERBS THAT DO WELL WHEN DIVIDED**
>
> bee balm
> chives
> coneflower
> lemon balm
> mint
> oregano
> parsley

This is a good method if you want to divide herbs from the garden for use in containers. Depending on the size of the division, the plants will mature and fill in a pot faster than plants produced by other propagation methods. You can also divide older plants when they overgrow their containers. Return a piece to the original container to refresh the space and use the remaining divisions elsewhere.

If you are dividing a container-grown plant and want to immediately put a smaller piece of it back in the same place rather than into a new pot, clean out the hole left behind from the removed portion. Tug out as much of the old root mass as possible, then loosen the soil. Add fresh potting soil to the open space and replant your division.

Above: With the proper equipment, dividing perennial herbs is easy.
Opposite: Fresh perennial herb divisions planted in 6-inch (15 cm) pots.

How to divide a container-grown perennial herb

Mint is one plant you must divide every year to keep it from overgrowing its container. Use this easy technique to divide mint, as well as most other perennial herbs.

Step 1: Carefully remove the plant from the container. To help carve into the soil, use a long blade pruning saw to remove as much of the plant and root ball as possible without disturbing other plants in the container. This can be tricky, but be patient. Reach down as far as you can to slice out deep-rooted perennials.

Step 2: Once removed, the plant, if it's not root-bound, should easily fall apart into small sections. If the root ball is tightly bound up, use a clean blade to gently score the roots around the outer edges to loosen them.

Step 3: Shake off as much soil as you can. Don't worry if some of the roots get a bit roughed up. With a little coaxing, you can usually pull apart or cut the plant. You are looking for a healthy set of roots attached to the base of each plant division.

Step 4: Once you have a section cleanly away, trim off any dark or broken roots or old stems. Remove any old growth, flowers, and yellowing or dead leaves.

Step 5: Immediately plant each division in a new pot and water it well. Never let the roots of your transplants sit in hot sun or dry out. They will quickly lose moisture and may not recover. Allow the division to grow in its new pot for a while, keeping it well watered. When you can tug on the plant and it does not easily come out of the pot, it is ready to go on display in your larger container gardens.

GROWING BY LAYERING

Layering is a good method to get new plants for use in containers from established herbs in the garden. Putting part of a branch in contact with the soil while it is still attached to the parent plant allows the branch to form roots and grow a small new plant right next to its parent. Once established, the plant can be cut from the main plant. Herbs that get large and woody are the best candidates for a layering technique.

EASY HERBS FOR LAYERING

lavender

mint

oregano

rosemary

thyme

Here's how to layer an herb plant:

Step 1: Choose a stem with a single, healthy branch hanging close to the soil.

Step 2: Without removing the branch from the parent plant, carefully strip the leaves from the bottom of the stem. Gently bend the stem until it is in contact with the ground.

Step 3: Secure the stem with a landscaping staple or cover it with a small mound of soil to keep it in contact with the soil. Rooting and new growth will take several months, so the stem must stay in contact with the ground.

Step 4: Once there is a small, established root system (check by tugging gently to test for resistance), cut the new plant from its parent and transfer it into a separate pot. Allow the plant to grow and fill out to a good size before moving it to a container garden.

RESOURCES

PLANTS AND SEEDS

Shop your local garden stores and farmers markets first!

Hudson Valley Seed Co.
(845) 204-8769
hudsonvalleyseed.com

Johnny's Selected Seeds
(877) 564-6697
johnnyseeds.com

Mountain Valley Growers
(559) 338-2775
mountainvalleygrowers.com

Renee's Garden Seeds
(888) 880-7228
reneesgarden.com

Richter's Herbs
(905) 640-6677
richters.com

Seed Savers Exchange
(563) 382-5990
seedsavers.org

Territorial Seed Company
(800) 626-0866
territorialseed.com

POTTERY AND CONTAINERS

Tractor Supply Co.
(galvanized containers)
tractorsupply.com

Pennoyer Newman
(high-end containers)
pennoyernewman.com

Veradek Outdoor
(metal and contemporary containers)
veradekshop.com

Whichford Pottery
(handmade terra cotta containers)
whichfordpottery.com

TOOLS, GROW LIGHTING, AND OTHER SUPPLIES

Gardener's Supply Company
gardeners.com

HERB PUBLICATIONS

Herb Quarterly
herbquarterly.com

Mother Earth News
motherearthnews.com

Mother Earth Living
motherearthliving.com

HERB SOCIETIES AND INFORMATION RESOURCES

American Botanical Council
herbalgram.org

Herb Research Foundation
herbs.org

Herb Society of America
herbsociety.org

International Herb Association
iherb.org

North Carolina State University
newcropsorganics.ces.ncsu.edu/herb

LES
FINES HERBES.

ACKNOWLEDGMENTS

As I ponder acknowledgments to a new book, I tend to take a stroll down memory lane of what brought me here. The photos and words bring me to places I visited and the seasons of my own garden. This book's bank of memories is no different. Years ago, I became a bit obsessed with how to make herbs more accessible to everybody, and container gardens do just that. More accessible means more useable, and that is what growing herbs is all about: the exploration, gardening, and use of plants to enrich our lives. Even if it's only one pot of aloe vera on the windowsill—there grows an herb garden.

Thank you Jessica Walliser for letting me run with the idea of this book. Thank you as well to the team at Cool Springs Press: Marissa Giambrone, Meredith Quinn, and all of the talents behind the scenes who I might never get to meet but have their fingerprints here somewhere. I appreciate the respect and opportunity to be an author in your world. An opportune editing moment came when I was reconnected with Lorraine Anderson. After almost a decade, I remembered why I love working with her on a project: Her gentle coaxing and garden knowledge makes editing painless—yes, really!

A big thank you to the talented Christina Salwitz for coming to play in my garden, ignoring the weeds, and emerging with all the "glamour" shots of container gardens. To garden owners who allowed me to capture photos and find inspiration in your spaces—David Perry, the Jamisons, Ravenna Gardens, Christiansen's, Zenith Holland, and the many community gardens and locations on my travels across the U.S. and Europe—thank you.

To clients and co-workers who gave me time and space to write this book: It is a rare and treasured gift to have someone say, "Get your book done. I'll wait for when you have time."

And of course, to friends, peers, and followers who love this garden life too: Thank you for all your encouragement and support.

And, yes, there is always room for another container in the garden!

ABOUT THE AUTHOR

Sue Goetz is an award-winning garden designer, writer, and speaker. Through her design business, she works with clients to personalize outdoor spaces, from garden coaching and container design to full-scale landscapes. Her garden design work has earned gold medals at the Northwest Flower and Garden Show and specialty awards from *Sunset* magazine, *Fine Gardening*, and the American Horticultural Society. Her motto, "Inspiring gardeners to create," defines her talks and hands-on workshops. She is the author of *The Herb Lover's Spa Book* and *A Taste for Herbs* and is a regular contributor to regional and national publications.

Sue is a certified professional horticulturist (CPH) and a sustainable landscape professional (Eco-pro). She has been named Educator of the Year (2012) by the Washington State Nursery and Landscape Association and shares her love of the garden and of herb growing all over the country. Sue lives in the beautiful Pacific Northwest, and when not up to her nose in herbs and dirt in the garden, she enjoys pen and ink botanical illustration and creating mixed media art with pressed plants.

Connect!
suegoetz.com
herbloversgarden.com
Facebook.com/CreativeGardener
Pinterest: Sue Goetz
Instagram: creativegardener
Twitter: @gardenersue
Linked In: Sue Goetz

INDEX

Note: Page numbers listed in italics indicate a photo is shown

Agastache. See Anise hyssop
Aging terra cotta pots, 18–19
Allium shoenoprasum (Chives), 77
Aloe vera (*aloe vera*), 48, 112, 166
Aloysia citriodora (lemon verbena). *See* Lemon verbena
Alpine strawberries, 32, 40
Anethum graveolens. See Dill
Angelica, 64
Anise hyssop
    attracting insects, 141
    'Blue Fortune', 62, 137, *139*, 141
    as deer resistant, 70
    having colorful foliage, 63
    for hummingbirds, 137
    propagating, 166
    seasonal color using, 62, 63
Apiaceae family, *141*
Areca palm (*Dypsis lutescens*), 45
Aromatherapy, 116–119
Artemisia, 54, 63, 166
Asters, 143
Australian bush mint, 115

Bagged soils, 150
Balconies, herb containers on, 35
Basil
    for aromatherapy, 119
    for cocktails, 104
    'Dark Opal', 63, 76, 94, 104
    as "filler" in a pot, 65
    'Genovese', 81
    grown in individual pots, 61
    harvesting, 159
    having colorful foliage, 62
    as healing herb, 112
    for living walls, 40
    microgreens, 94
    for mixed plantings, 60
    'Mrs. Burns Lemon', 123
    pot depth for, 29
    propagating, 166
    'Red Rubin', 75–76, 94, 104
    seed collecting from, 169
    in Simple Syrup recipe, 107
    Thai, *34*, 89, 91, 119
    Tulsi (holy basil), 91
    used for garnish, 104
    as warm-season herb, 148
    for window boxes, 42
Bay laurel (*Laurus nobilis*), 48, 50, 66, 67, 79, 82, 85, 166
Beach glass, as topdressing, 154
Bee balm
    for bees and butterflies, 143
    growing by division, 176
    for hummingbirds, 137

    as lemony, 123
    powdery mildew and, 157
    propagating, 166
    seasonal color using, 62
    for shade, 32
Bees
    gathering dish for, 144
    herbs for, 143
Beverages, 97–107
    cocktails, 102–107
    tea, 99–101
Biennial seeds, 168
Black-eyed Susan, 143, *143*
Blueberry, Jelly Bean, 104–105
Blue blooms, plants with, 62
Blue foliage, plants with, 63
'Blue Fortune' anise hyssop, 62, 137, *139*, 141
Bok choy (pak choi), 91
Bold-colored pottery, 56, *57*
Borage, *60*
    for bees and butterflies, 143
    grown in individual pots, 61
    for herbal cocktail, *103*
    microgreens, 94
    propagating, 166
    seasonal color using, 62
    used for garnish, 104
Botanical names, 50
Boxwood, 115
Bronze fennel, 63
Bronze foliage, plants with, 63
Buddleia (butterfly bush), 143
Bulb fennel, 87
Butterflies, herbs for, 143
Butterfly bush, 143

Calendula, 10, *161*
    as deer resistant, 70
    grown in individual pots, 61
    harvesting, 159
    as healing herb, 112
    for mixed plantings, 60
    for natural beauty, 129
    planting with rosemary, 160
    pot depth for, 29
    propagating, 166
    seasonal color using, 62, 111
    in window boxes, 42
Camellia sinensis (Tea plant), 9, 49, 68, 99, 101, 168
Cardoon, 64, *64*
Carolina allspice (*Calycanthus* spp.), 67
Cast stone pots, 20
Catmint, *34*
    attracting pollinators, 141, 142
    as deer resistant, 70

for hanging baskets, 41
for hummingbird gardens, 139
as insect repellent, 123
'Junior Walker', 141, 142
planting with lavender, 161
propagating, 166
seasonal color using, 62
Ceramic pots, 15–20
Chamaemelum nobile. See Chamomile
Chamomile, *21, 127*
for aromatherapy, 117, 119
for hanging baskets, 41
as healing herb, 112
for natural beauty, 129
propagating, 166
seasonal color using, 62
as "spiller" in a pot, 64
for tea, 99, 172
used for garnish, 104
Chaste tree (*Vitex* spp.), 68
Chervil, 57, 87, 148, 166
Chimney flues, as containers, 75
Chives (*Allium schoenoprasum*), 76
for bees and butterflies, 143
as cool-season herb, 148
as deer resistant, 70
growing, 76
growing by division, 176
growing from seed, 167
pot depth for, 29
propagating, 166
seasonal color using, 62
seed collecting from, 169
for shade, 32
used for garnish, 104
for window boxes, 42
Choisya ternata, 123
Cilantro, *60*
as cool-season herb, 148
as culinary herb, 79
as fill-in plant, 161
grown in individual pots, 61
microgreens, 94
for mixed plantings, 60, *60*
propagating, 16
for shade, 32
for Thai cooking, 89
Citrus, 92–93, 103
Citrus x *latifolia* (Lime, Persian), 89–90
Clay pots, 153
Cleaning indoor herbs, 46
Cleanser, herbal, 133
Coir, in soil mix, 151
Color(s)
of containers, 56–57
flowering herbs for seasonal, 62
foliage, 63
that attract hummingbirds, 139
Columnar apple trees, 143
Common names, 50
Compost, in soil mix, 151

Concrete pots, 21
Coneflowers, purple. *See* Purple coneflower (*Echinacea purpurea*)
Containers. *See also* Pots
chimney flues as, 75
color of, 56–57
connecting with your home's style, 54
grouping and arranging, 66–70
grow bags, 23
reasons for growing herbs in, 10–11
soil for, 149–151
for starting seeds, 170–171
Cool colors, 57
Cool-season herbs, 148
Coriandrum sativum. See Cilantro
Crape myrtle (*Lagerstroemia*), 143
Crassula spp. (jade plant), 45
Creeping oregano, 40
Creeping rosemary, *56*
Creeping sweet marjoram, 40
Crocus, 143
Crocus sativus (saffron), 42, 93
Croton, 45
Culinary herbs, 73–95
harvesting, 73
for microgreens, 94–95
profiles of French, 84–87
profiles of Italian, 80–83
profiles of Thai, 88–91
profiles of traditional, 75–79
rare and unusual, 92–93
Curly parsley, *9, 24, 58, 163*
for fill-in around seasonal herbs, 161
flat-leaf parsley *vs.*, 77
for living walls, 40
used for garnish, 104
for window boxes, 42
Curry plant, 63
Cuttings, growing herbs by, 172–175
Cymbopogon citratus (lemongrass), 89, 90, 121, 122, *123,* 133, 161, 166

Daffodil, 143
Damping off, 172
'Dark Opal' basil, 63, 76, 94, 104
Decks, herb containers on, 38
Deer-resistant herbs, 70
Dill, *167*
for bees and butterflies, 143
'Bouquet', 75, 76
as cool-season herb, 148
grown in their own pot, 28
microgreen, 94
in mixed plantings, 60
propagating, 166
seed collecting from, 169
Disease, 157, 172
Division, growing herbs by, 176–178
Drainage, 17, 45, 149, 151
Drought-tolerant herbs, 33, 153
Dwarf catmint, 41

Dwarf fruit trees, 143
Dwarf ginkgo, 114, *114*

Earthenware, 20
Echinacea purpurea (purple coneflower), 50, *69*, 138,
        142, 166
Echinacea 'White Swan', 62
Elderberry, 56, *69*
        colorful foliage of, 63
        propagating, 166
        as "thriller" in container garden, 64
        varieties, 68
Elderflower
        seasonal color using, 62
English lavenders, *34*, 99–100, 105, 111, 127–128, 131, 142,
        143, 161
        for bees and butterflies, 143
        'Munstead', 131
English thyme, 132–133
Entryway, herb containers near, 36

Fall, container garden chores during, 163
Fennel, *25, 34, 54*
        attracting insects, 141
        as cool-season herb, 148
        as fill-in plant, 161
        grown in their own pot, 28
        in Italian culinary garden, 82
        microgreens, 94
        in mixed plantings, 60
        pot depth for, 29
        propagating, 166
        seasonal color with, 63
        seed collecting from, 169
        as "thriller" in a pot, 64
        types of, 87
Fertilizer/fertilizing, 46, 153
Fiberglass containers, 24
Flat-leaf parsley
        in Broiled Herb and Cheese Tomatoes, 83
        growing, 77, 82
        used for garnish, 104
Florence fennel, 87
Flowering herbs, 62, 103, 121
Flues, chimney, 75
Foeniculum vulgare. See Fennel
Foot soak, herbal, 129
French culinary herbs, 84–87
French lavender, 118, 127–128
French marigolds, 157
French tarragon
        for the culinary garden, 79, 86
        as deer resistant, 70
        for living walls, 40
        propagating, 166, 172
        for shade, 32
        as warm-season herb, 148
        for window boxes, 42

Galium odoratum. See Sweet woodruff
Galvanized metal containers, 22

Garden beds, herb containers in, 34
Garlic chives
        for Asian flavor, 91
        growing from seed, 167
        for living walls, 40
        propagating, 166
        seasonal color using, 62
Garnishes, herbs for, 104
'Genovese' basil, 81, 94
Geranium 'Biokovo', 123
Geranium leaves
        in foot soak, 129
        in potpourri, 117
Germander, 65, 115
Germination, 95, 169, 171
Ginger (*Zingiber officinale*), 92
Ginkgo (*ginkgo biloba*), 68
        'Mariken', 111
        medicinal uses, 111
        propagating, 166
Globe artichoke, 64
Goldenrod, 143
Gold foliage, plants with, 63
Gravel
        at bottom of pot, 149
        as topdressing, 154
Grow bags, 23, *36*
Grow lights, 45

Hamamelis spp., 9
Hanging baskets, 38–42, 150
Hardening plants, 171–172
Hardy fuchsia, 143
Harvesting herbs, 99, 159–160
Hazelnut shells, 154
Healing herbs, 109–123
Hedychiums (hardy ginger), 92
Herbal trees and shrubs. *See* Trees and shrubs
Herb container gardens. *See also* Containers; Herbs
        composing groupings of herbs in, 57, 59–61
        factors related to location of, 32–33
        fill-in plants for, 161
        fragrance from, and placement of, 117
        indoors, 43–50
        outdoor locations for, 34–42
        planning for succession, 160–161
        seasonal calendar for, 162–163
        tips for keeping tidy, 161
Herbes de Provence, 87
Herbs. *See also* Names of individual herbs
        attracting hummingbirds, 136–139
        for bees and butterflies, 143
        botanical definition of, 9
        choosing what to grow, 11–12
        for cocktails, 102–107
        with colorful foliage, 63
        culinary. *See* Culinary herbs
        deer-resistant, 70
        flowering, 62
        for garnishes, 104
        to grow indoors, 48–49

lemon, 121–123
medicinal, 111–113
shopping for, 147–148
for tea, 99–101
that are thrillers, spillers, and fillers, 64–65
topiary, 114–115
used for aromatherapy, 116–119
used for house cleaning, 130–133
used for skin care products, 127–129
Hibiscus, 45
Hot colors, 56
House cleaning, herbs used for, 130–133
Humidity, 46
Hummingbirds, herbs attracting, 136–139
Hyacinth, 143
Hyssop. *See* Anise hyssop

Insect-repellant plants, 123
Insects, herbs attracting beneficial, 141–144
Italian culinary herbs, 80–83
Italian parsley. *See* Flat-leaf parsley

Jade plants (*Crassula* spp.), 45
Japanese mock orange, 123
Jewel tones, herbs with, 57

Lady's mantle, seasonal color using, 62
Laurus nobilis. *See* Bay laurel (*Laurus nobilis*)
Lavandula. *See* Lavender
Lavandula x *intermedia*, 127
Lavender
'Anouk', 141, 142
attracting pollinators, 142
as deer resistant, 70
growing by layering, 179
growing from cuttings, 175
harvesting, 160
having colorful foliage, 63
'Hidcote', 99–100, 103, 105, 111
medicinal uses, 111
'Munstead', 131
planting with catmint, 161
propagating, 166
'Royal Velvet', 127
seasonal color using, 62
in Simple Syrup recipe, 107
for skin care products, 127–128
Spanish, 115, 118, 142
used for garnish, 104
as warm-season herb, 148
Lavender buds
in foot soak, 129
for midsummer healing tea, 113
in potpourri, 117
in Simple Syrup, 107
Layering, growing herbs by, 179
LED grow light, 45
Lemon balm, 123
for bees and butterflies, 143
as deer resistant, 70
growing by division, 176

grown in their own pot, 28
medicinal uses, 112, 113
microgreens, 94
in midsummer healing tea, 113
for natural beauty, 129
propagating, 166
for shade, 32
for tea, 100, 113
Lemon fragrance, trees and shrubs with, 123
Lemongrass (*Cymbopogan citratus*), 89, 90, 121, 122, *123*, 133, 161, 166
Lemon herbs, 121–123
Lemon thyme (*Thymus* x *citriodorus*), 78–79, 117, 119, 122
'Lemon Twist' calendula, 121
Lemon verbena, *61*
care of, 100, 122
for cocktail container garden, 106
growing from cuttings, 175
for household cleaning, 131–132
as houseplant, 48
propagating, 166
Light requirements
indoor herb containers, 44–45
location of container herbs and, 32
microclimate and, 33
when growing from seed, 171
Lime, Persian, 89–90
Living walls, herb containers for, 38–41
Lovage, 32, 64, 166

Magnolia grandiflora, 123
Manure, in soil mix, 151
Marigold. *See* Calendula
Marjoram
creeping sweet, for living walls, 40
for culinary garden, 82, 85–86
harvesting, 159
propagating, 166
Medicinal herbs, 109–123
Melissa officinalis. *See* Lemon balm
Mentha spicata (spearmint), 100, 106
Mentha spp. *See* Mint(s)
Mentha suaveolens 'Variegata' (pineapple mint), 63, 103
Mentha x *piperita*. *See* Peppermint (*Mentha* x *piperita*)
Mentha x *piperita* 'Grapefruit', 138
Metal containers, 22
Mexican orange, 123
Meyer lemon, 103
Microclimates, 33
Microgreens, 94–95
Mint(s). *See also* Peppermint (*Mentha* x *piperita*);
Spearmint (*Mentha spicata*)
care of, 49
as cool-season herb, 148
as deer resistant, 70
as "filler" in a pot, 65
'Grapefruit', 138
growing by division, 176, 178
growing by layering, 179
growing from cuttings, 172, 173
grown in their own pot, 28

harvesting, 159
as healing herb, 112
for living walls, 40
propagating, 166
Mint (lemon), 123
Mixed plantings, 60
Mock orange, 123
Monarda spp. (bee balm). *See* Bee balm
'Mrs. Burns Lemon' basil, 123
Myrtle, 115

Nasturtiums, *16*
for balconies, 35
as fill-in plant, 161
grown in individual pots, 61
for hanging baskets, 41
for hummingbird gardens, 139
for living walls, 40
planting with rosemary, 160
propagating, 166
used for garnish, 104
for window boxes, 42
Ninebark (*Physocarpus*), 143

Ocimum basilicum. See Basil
Orange blooms, plants with, 62
Oregano
attracting pollinators, 142
Greek, 77
growing by division, 176
growing by layering, 179
grown in their own pot, 28
harvesting, 159
having colorful foliage, 63
'Hot and Spicy', 82
'Kent Beauty', 142
propagating, 166
seasonal color using, 62
as "spiller" in a pot, 64
as warm-season herb, 148
Origanum. See Oregano
Origanum majorana. See Marjoram
Outdoors, moving indoor herbs, 50

Pansies, 104, 143
Parsley, *44. See also* Curly parsley
for bees and butterflies, 143
as cool-season herb, 148
'Extra Curled Dwarf', 128
as "filler" in a pot, 65
flat-leaf, 77, 82, 83, 104
growing by division, 176
microgreens, 94
for mixed plantings, 60
pot depth for, 29
propagating, 166
for shade, 32
for skin care, 128
Patchouli, 119, 166
Patios, herb containers on, 38
Peat pellets, 170, *170*

Pelargonium 'Citrosum', 123
Pelargonium spp. *See* Scented geraniums
Pennyroyal, for hanging baskets, 41
Peppermint (*Mentha* x *piperita*), 32, 75, 78, 119, 133, *152, 175*
'Peppermint', scented geranium, 117, 118, 131, 132
Perennial seeds, 168
Perilla frutescens. See Shiso
Perlite, in soil mix, 151
Persian lime, 89–90
Pests, 157
Petroselinum. See Parsley
Philadelphus spp. (mock orange), 123
Physocarpus (ninebark), 143
Pineapple mint, 63, 103
Pineapple sage, 100, 137, 139
Pittosporum tobira, 123
Plain terra cotta pots, 15–16
Plastic containers, 24
Pollinator herbs, 135–144
Portable pot template, 59
Pot marigold. *See* Calendula
Potpourri, 117, 118, 119
Pots. *See also* Containers
cachepots for, 29
concrete and cast stone, 21
considerations when choosing, 27–29
filling, 149–151
glazed ceramic, 20
metal, 22
plastic and fiberglass, 24
portable template of, 59
soil volume for, 150
terra cotta, 15–19
vintage and repurposed, 26–27
wood, 24–25
Powdery mildew, 137, 157
Propagation, 165–179
checklist, by plant, 166
by cuttings, 172–175
by division, 176–178
by layering, 179
by seed, 167–172
Prostanthera (Australian bush mint), 115
Purple blooms, plants with, 62
Purple coneflower (*Echinacea purpurea*), 50, 62, *69*, 138, 142, 166, 176
Purple foliage, plants with, 63
Purple leaf basils, 104
Purple Shiso. *See* Shiso

Raspberry, Raspberry Shortcake (*Rubus idaeus* 'NR7'), 106
Red blooms, plants with, 62
Red castor bean, 64
'Red Rubin' basil, 75–76, 94, 104
Repotting, 154
Rhizomes, of ginger, 92
Rocks
at bottom of pot, 149
as topdressing, 154
Rooftops, herb containers on, 36–37

Root-bound plants, 82, 148, 154, 156, 171
Rosemary
    for aromatherapy, 119
    'Arp', 65, 131, 132
    'Barbeque', 82, 138
    for bees and butterflies, 143
    for cocktail container garden, 106
    as deer resistant, 70
    as "filler" in a pot, 65
    growing by layering, 179
    growing from cuttings, 175
    for hanging baskets, 41
    harvesting, 159
    'Irene', 112
    for living walls, 40
    medicinal uses, 112
    in midsummer healing tea, 113
    as a pollinator, 138
    propagating, 166
    as "spiller" in a pot, 64
    used for garnish, 104
    used for household cleaning, 132
    as warm-season herb, 148
Rose petals, in potpourri, 117
Rose-scented geranium, *60*, 100, 118, 127, 128
Roses/rose shrub
    for aromatherapy, 117, 118
    growing, 118
    harvesting, 159
    for hummingbird garden, 139
    propagating, 166
    seasonal color using, 62
    for tea, 100
Rosmarinus officinalis. See Rosemary
Rue
    as deer resistant, 70
    as "filler" in a pot, 65
    having colorful foliage, 63
    silver and blue foliage, 63

Saffron (*Crocus sativus*), 42, 93, 166
Sage, *44, 53*
    'Berggarten', 70, 85, 86, 132
    for cocktail container garden, 106
    as culinary herb, 79, 82
    as deer resistant, 70
    'Golden Delicious' (pineapple sage), 100, 137, 139
    growing from cuttings, 175
    harvesting, 159
    having colorful foliage, 63
    as healing herb, 112
    'Icterina', 63
    for natural beauty, 129
    planted in individual pots, 61
    as a pollinator, 138
    propagating, 166
    'Purpurascens', 63
    'Purpurea', 138
    in Salvias family, 83
    seasonal color with, 63
    'Tricolor', 63

    used for garnish, 104
    used for household cleaning, 132
    as warm-season herb, 148
Salvia. See Sage
Sand, in soil mix, 151
Santolina
    as "filler" in a pot, 65
    having colorful foliage, 63
    propagating, 166
Satureja hortensis (summer savory), 78, 79
Satureja montana (savory, winter), 41, 78, 159
Savory
    as culinary herb, 78
    for hanging baskets, 41
    harvesting, 159
    propagating, 166
    summer, 78, 79
    used for garnish, 104
    winter, 41, 78, 159
Scented geraniums
    for aromatherapy, 118–119
    'Chocolate Mint', 63
    'Citronella', 123
    growing from cuttings, 172, *173*, 175
    having colorful foliage, 63
    for houseplants, 49
    'Lady Plymouth', 63, 103, 106
    'Lady Plymouth Variegated', 128–129
    'Lemon Fizz', 121, 122
    'Mabel Grey', 123
    'Old Fashioned Rose', 118–119
    'Peppermint', 118, 131, 132
    propagating, 166
    rose. *See* Rose-scented geranium
    for skin care, 128–129
    as topiary, 115
    used for household cleaning, 132
Seed, growing herbs from, 167–172
Seedlings
    chamomile tea for fungal issues with, 172
    hardening, 171–172
    transplanting into containers, 171
    when to plant in containers, 170
Self-sowing plants, 169
Shade
    hanging baskets for, 41
    herbs for, 32, 41
Shiso, 57, 75, 78, 91, 94, 161, 166
Shrubs. *See* Trees and shrubs
Signet marigold, 166
Silver foliage, plants with, 163
Simple syrup recipe, 107
Skin care products, herbs for, 127–129
Smudge sticks, 133
Snake plant (*Sansevieria trifasciata*), 45
Soil
    for potted gardens, 149–151
    replenishing, 154
    rooting cuttings in, 174
    for seed starting, 171
Soil mix, 151

Soil volume, 150
Southern magnolia, 123
Spanish lavender, 54, 115, 118, 142
Spearmint (*Mentha spicata*), 100, 104, 106
Sphagnum peat moss, in soil mix, 151
Spring, container garden chores during, 162
Steps, herb containers on, 36
Stevia
    for cocktail container garden, 106
    as houseplant, 49
    for mixed plantings, 60
    propagating, 166
    for tea, 100
    as warm-season herb, 148
Stevia rebaudiana. See Stevia
Summer, container garden chores during, 162
Summer savory (*Satureja hortensis*), 78, 79
Sweet woodruff, *64*
    for hanging baskets, 41
    seasonal color using, 62
    for shade, 32
    as "spiller" in a pot, 64
    for tea, 101

Tea(s), 99–101, 113, 172
Tea plant (*Camellia sinensis*), 9, 49, 68, 99, 101, 166
Temperature
    for coddling cuttings, 174
    for growing from seed, 171
    for indoor herb containers, 45
Template, portable pot, 59
Terra cotta pieces, as topdressing, 154
Terra cotta pots, 15–19
Thai basil, *34*, 91, 119
Thai culinary herbs, 88–91
Thrillers, spillers, and fillers formula, 64–65
Thyme
    for aromatherapy, 119
    for bees and butterflies, 143
    for cocktail container garden, 106
    as deer resistant, 70
    English, 132–133
    as "filler" in a pot, 65
    'Foxley', 119, 138
    growing by layering, 179
    for hanging baskets, 41
    harvesting, 159
    as having colorful foliage, 63
    as healing herb, 112
    lemon, 78–79, 119, 122
    for living walls, 40
    as a pollinator, 138
    propagating, 166
    'Silver Posie', 63, 131, 133
    as "spiller" in a pot, 64
    used for household cleaning, 132–133
    variegated lemon, 90, 119, 122

Thymus pulegiodes 'Foxley', 119, 138
Thymus vulgaris, 132–133
Thymus x *citriodorus* (lemon thyme), 78–79, 90, 119, 122
Topdressing, 153–154, *155*
Topiary herbs, 114–115
Trees and shrubs, 9
    insect-attracting, 143
    with a lemony fragrance, 123
    pot depth for, 29
    used for architecture, 67–68
Trellis, 33
Tulsi (holy basil), 91

Umbelliferae, *141*

Variegated lemon thyme, 119, 122
Variegated mint, 41, 73, 104
Vermiculite, in soil mix, 151
Vertical planting walls, 38–40
Vintage containers, 26
Violas
    for balconies, 35
    as cool-season herb, 148
    encouraging pollinators, 143
    grown in individual pots, 61
    for hanging baskets, 41
    for living walls, 40
    for pollinators, 143
    seasonal color using, 62
    for shade, 32
    used for garnish, 104
Violets, 32, 161

Warm-season herbs, 148
Watering, 153
    considered when choosing where to grow herbs, 32
    for hanging baskets and living walls, 42
    for indoor herb containers, 46
    location of container herbs and, 32
    when growing from seed, 171
    when using terra cotta pots, 17
    for window boxes, 42–43
Water, rooting cuttings in, 175
White blooms, plants with, 62
Window boxes
    herbs grown in, 42–43
    soil volume for, 150
Winter
    care for terra cotta plants during, 17
    container garden chores during, 163
Winter daphne (*Daphne odora*), 123
Winter savory (*Satureja montana*), 41, 78, 159
Witch hazel (*Hamamelis* spp.), 9, 67, 68, 128

Yellow blooms, plants with, 62
Yellow lavender, 123

Zingiber officinale (ginger), 92